CLEAN WATER

EARTH • AT • RISK

Acid Rain

Alternative Sources of Energy

Animal Welfare

The Automobile and the Environment

Clean Air

Clean Water

Degradation of the Land

Economics and the Environment

Environmental Action Groups

Environmental Disasters

The Environment and the Law

Extinction

The Fragile Earth

Global Warming

The Living Ocean

Nuclear Energy • Nuclear Waste

Overpopulation

The Ozone Layer

The Rainforest

Recycling

Solar Energy

Toxic Materials

What You Can Do for the Environment

Wilderness Preservation

CLEAN WATER

by Karen Barss

Introduction by
Russell E. Train

Chairman of
the Board of Directors,
World Wildlife Fund and
The Conservation Foundation

CHELSEA HOUSE PUBLISHERS

new york   philadelphia

CHELSEA HOUSE PUBLISHERS
EDITOR-IN-CHIEF: Remmel Nunn
MANAGING EDITOR: Karyn Gullen Browne
COPY CHIEF: Mark Rifkin
PICTURE EDITOR: Adrian G. Allen
ART DIRECTOR: Maria Epes
ASSISTANT ART DIRECTOR: Howard Brotman
MANUFACTURING DIRECTOR: Gerald Levine
SYSTEMS MANAGER: Lindsey Ottman
PRODUCTION MANAGER: Joseph Romano
PRODUCTION COORDINATOR: Marie Claire Cebrián

EARTH AT RISK
Senior Editor: Jake Goldberg

Staff for *Clean Water*
COPY EDITOR: Ian Wilker
EDITORIAL ASSISTANT: Danielle Janusz
PICTURE RESEARCHER: Villette Harris
DESIGNER: Maria Epes
LAYOUT: Marjorie Zaum
COVER ILLUSTRATION: Danny O'Leary

5 7 9 8 6 4

Library of Congress Cataloging-in-Publication Data
Barss, Karen.
  Clean water/Karen Barss; introduction by Russell E. Train.
    p. cm.—(Earth at risk)
  Includes bibliographical references and index.
  Summary: Discusses the problem of maintaining a clean water
supply and relates this issue to such topics as pollution, depletion
of sources, and other environmental concerns.
  ISBN 0-7910-1583-1
       0-7910-1608-0 (pbk.)
  1. Water quality—Juvenile literature. 2. Water—
Pollution—Juvenile literature. [1. Water quality
management. 2. Water—Pollution. 3. Pollution.]
I. Title II. Series   91-24149
TD370.B37 1992   CIP
363.6'1—dc20   AC

C O N T E N T S

Introduction—Russell E. Train   6

1 The Clean Water Problem   13

2 Water Resources   25

3 Threats to the Water Supply   35

4 Water and Agriculture   45

5 Water and Industry   59

6 Municipal Water Use   71

7 Government Action   83

8 Protecting the Future   93

Appendix: For More Information   100

Further Reading   102

Glossary   104

Index   106

Conversion Table   111

INTRODUCTION

Russell E. Train

Administrator, Environmental Protection Agency, 1973 to 1977; Chairman of the Board of Directors, World Wildlife Fund and The Conservation Foundation

There is a growing realization that human activities increasingly are threatening the health of the natural systems that make life possible on this planet. Humankind has the power to alter nature fundamentally, perhaps irreversibly.

This stark reality was dramatized in January 1989 when *Time* magazine named Earth the "Planet of the Year." In the same year, the Exxon *Valdez* disaster sparked public concern over the effects of human activity on vulnerable ecosystems when a thick blanket of crude oil coated the shores and wildlife of Prince William Sound in Alaska. And, no doubt, the 20th anniversary celebration of Earth Day in April 1990 renewed broad public interest in environmental issues still further. It is no accident then that many people are calling the years between 1990 and 2000 the "Decade of the Environment."

And this is not merely a case of media hype, for the 1990s will truly be a time when the people of the planet Earth learn the meaning of the phrase "everything is connected to everything else" in the natural and man-made systems that sustain our lives. This will be a period when more people will understand that burning a tree in Amazonia adversely affects the global atmosphere just as much as the exhaust from the cars that fill our streets and expressways.

Central to our understanding of environmental issues is the need to recognize the complexity of the problems we face and the relationships between environmental and other needs in our society. Global warming provides an instructive example. Controlling emissions of carbon dioxide, the principal greenhouse gas, will involve efforts to reduce the use of fossil fuels to generate electricity. Such a reduction will include energy conservation and the promotion of alternative energy sources, such as nuclear and solar power.

The automobile contributes significantly to the problem. We have the choice of switching to more energy efficient autos and, in the longer run, of choosing alternative automotive power systems and relying more on mass transit. This will require different patterns of land use and development, patterns that are less transportation and energy intensive.

In agriculture, rice paddies and cattle are major sources of greenhouse gases. Recent experiments suggest that universally used nitrogen fertilizers may inhibit the ability of natural soil organisms to take up methane, thus contributing tremendously to the atmospheric loading of that gas—one of the major culprits in the global warming scenario.

As one explores the various parameters of today's pressing environmental challenges, it is possible to identify some areas where we have made some progress. We have taken important steps to control gross pollution over the past two decades. What I find particularly encouraging is the growing environmental consciousness and activism by today's youth. In many communities across the country, young people are working together to take their environmental awareness out of the classroom and apply it to everyday problems. Successful recycling and tree-planting projects have been launched as a result of these budding environmentalists who have committed themselves to a cleaner environment. Citizen action, activated by youthful enthusiasm, was largely responsible for the fast-food industry's switch from rainforest to domestic beef, for pledges from important companies in the tuna industry to use fishing techniques that would not harm dolphins, and for the recent announcement by the McDonald's Corporation to phase out polystyrene "clam shell" hamburger containers.

Despite these successes, much remains to be done if we are to make ours a truly healthy environment. Even a short list of persistent issues includes problems such as acid rain, ground-level ozone and smog, and airborne toxins; groundwater protection and nonpoint sources of pollution, such as runoff from farms and city streets; wetlands protection; hazardous waste dumps; and solid waste disposal, waste minimization, and recycling.

Similarly, there is an unfinished agenda in the natural resources area: effective implementation of newly adopted management plans for national forests; strengthening the wildlife refuge system; national park management, including addressing the growing pressure of development on lands surrounding the parks; implementation of the Endangered Species Act; wildlife trade problems, such as that involving elephant ivory; and ensuring adequate sustained funding for these efforts at all levels of government. All of these issues are before us today; most will continue in one form or another through the year 2000.

Each of these challenges to environmental quality and our health requires a response that recognizes the complex nature of the problem. Narrowly conceived solutions will not achieve lasting results. Often it seems that when we grab hold of one part of the environmental balloon, an unsightly and threatening bulge appears somewhere else.

The higher environmental issues arise on the national agenda, the more important it is that we are armed with the best possible knowledge of the economic costs of undertaking particular environmental programs and the costs associated with not undertaking them. Our society is not blessed with unlimited resources, and tough choices are going to have to be made. These should be informed choices.

All too often, environmental objectives are seen as at cross-purposes with other considerations vital to our society. Thus, environmental protection is often viewed as being in conflict with economic growth, with energy needs, with agricultural productions, and so on. The time has come when environmental considerations must be fully integrated into every nation's priorities.

One area that merits full legislative attention is energy efficiency. The United States is one of the least energy efficient of all the industrialized nations. Japan, for example, uses far less energy per unit of gross national product than the United States does. Of course, a country as large as the United States requires large amounts of energy for transportation. However, there is still a substantial amount of excess energy used, and this excess constitutes waste. More fuel efficient autos and home heating systems would save millions of barrels of oil, or their equivalent, each year. And air pollutants, including greenhouse gases, could be significantly reduced by increased efficiency in industry.

I suspect that the environmental problem that comes closest to home for most of us is the problem of what to do with trash. All over the world, communities are wrestling with the problem of waste disposal. Landfill sites are rapidly filling to capacity. No one wants a trash and garbage dump near home. As William Ruckelshaus, former EPA administrator and now in the waste management business, puts it, "Everyone wants you to pick up the garbage and no one wants you to put it down!"

At the present time, solid waste programs emphasize the regulation of disposal, setting standards for landfills, and so forth. In the decade ahead, we must shift our emphasis from regulating waste disposal to an overall reduction in its volume. We must look at the entire waste stream, including product design and packaging. We must avoid creating waste in the first place. To the greatest extent possible, we should then recycle any waste that is produced. I believe that, while most of us enjoy our comfortable way of life and have no desire to change things, we also know in our hearts that our "disposable society" has allowed us to become pretty soft.

Land use is another domestic issue that might well attract legislative attention by the year 2000. All across the United States, communities are grappling with the problem of growth. All too often, growth imposes high costs on the environment—the pollution of aquifers; the destruction of wetlands; the crowding of shorelines; the loss of wildlife habitat; and the loss of those special places, such as a historic structure or area, that give a community a sense of identity. It is worth noting that growth is not only the product of economic development but of population movement. By the year 2010, for example, experts predict that 75% of all Americans will live within 50 miles of a coast.

It is important to keep in mind that we are all made vulnerable by environmental problems that cross international borders. Of course, the most critical global conservation problems are the destruction of tropical forests and the consequent loss of their biological capital. Some scientists have calculated extinction rates as high as 11 species per hour. All agree that the loss of species has never been greater than at the present time; not even the disappearance of the dinosaurs can compare to today's rate of extinction.

In addition to species extinctions, the loss of tropical forests may represent as much as 20% of the total carbon dioxide loadings to the atmosphere. Clearly, any international approach to the problem of global warming must include major efforts to stop the destruction of forests and to manage those that remain on a renewable basis. Debt for nature swaps, which the World Wildlife Fund has pioneered in Costa Rica, Ecuador, Madagascar, and the Philippines, provide a useful mechanism for promoting such conservation objectives.

Global environmental issues inevitably will become the principal focus in international relations. But the single overriding issue facing the world community today is how to achieve a sustainable balance between growing human populations and the earth's natural systems. If you travel as frequently as I do in the developing countries of Latin America, Africa, and Asia, it is hard to escape the reality that expanding human populations are seriously weakening the earth's resource base. Rampant deforestation, eroding soils, spreading deserts, loss of biological diversity, the destruction of fisheries, and polluted and degraded urban environments threaten to spread environmental impoverishment, particularly in the tropics, where human population growth is greatest.

It is important to recognize that environmental degradation and human poverty are closely linked. Impoverished people desperate for land on which to grow crops or graze cattle are destroying forests and overgrazing even more marginal land. These people become trapped in a vicious downward spiral. They have little choice but to continue to overexploit the weakened resources available to them. Continued abuse of these lands only diminishes their productivity. Throughout the developing world, alarming amounts of land rendered useless by over-grazing and poor agricultural practices have become virtual wastelands, yet human numbers continue to multiply in these areas.

From Bangladesh to Haiti, we are confronted with an increasing number of ecological basket cases. In the Philippines, a traditional focus of U.S. interest, environmental devastation is widespread as defores-tation, soil erosion, and the destruction of coral reefs and fisheries combine with the highest population growth rate in Southeast Asia.

Controlling human population growth is the key factor in the environmental equation. World population is expected to at least double to about 11 billion before leveling off. Most of this growth will occur in the poorest nations of the developing world. I would hope that the United States will once again become a strong advocate of international efforts to promote family planning. Bringing human populations into a sustainable balance with their natural resource base must be a vital objective of U.S. foreign policy.

Foreign economic assistance, the program of the Agency for International Development (AID), can become a potentially powerful tool for arresting environmental deterioration in developing countries. People who profess to care about global environmental problems— the loss of biological diversity, the destruction of tropical forests, the greenhouse effect, the impoverishment of the marine environment, and so on—should be strong supporters of foreign aid planning and the principles of sustainable development urged by the World Commission on Environment and Development, the "Brundtland Commission."

If sustainability is to be the underlying element of overseas assistance programs, so too must it be a guiding principle in people's practices at home. Too often we think of sustainable development only in terms of the resources of other countries. We have much that we can and should be doing to promote long-term sustainability in our own resource management. The conflict over our own rainforests, the old growth forests of the Pacific Northwest, illustrates this point.

The decade ahead will be a time of great activity on the environmental front, both globally and domestically. I sincerely believe we will be tested as we have been only in times of war and during the Great Depression. We must set goals for the year 2000 that will challenge both the American people and the world community.

Despite the complexities ahead, I remain an optimist. I am confident that if we collectively commit ourselves to a clean, healthy environment we can surpass the achievements of the 1980s and meet the serious challenges that face us in the coming decades. I hope that today's students will recognize their significant role in and responsibility for bringing about change and will rise to the occasion to improve the quality of our global environment.

This picturesque freshwater stream is just part of the hydrologic cycle. Streams run into rivers, which run into the sea, from which water evaporates into the atmosphere. The evaporated water forms clouds and eventually precipitation, which replenishes streams.

chapter 1

THE CLEAN WATER PROBLEM

Water is our most precious natural resource. People can live without oil or gold or natural gas, but they can survive only a few days without water. Water has many uses besides simply quenching our thirst. At home, water cleans dishes, clothes, and people. It cooks food, delivers heat to radiators, and carries away sewage. It plays many roles in industrial processes and provides energy through hydropower. It is essential to the growth of plants and, therefore, to the food supply. It supports populations of birds, fish, aquatic plants, and mammals. It offers an alternative to land and air travel that is essential for large cargo. And it has many recreational uses, as well as aesthetic value, in city fountains and woodland streams. Yet for all its vital roles, water is often taken for granted, in part because it costs so little and the supply seems limitless. People rarely question whether water will flow when they turn on the tap.

This book considers the results of such a casual attitude toward water, in terms of both overusing the supply and polluting what is left. Oceans cover 70% of the earth's surface. That may seem like an infinite supply of water, but only 3% of the earth's water is fresh and suitable for drinking, crop irrigation, and most other uses. Seventy-five percent of that relatively small amount of fresh water is frozen in glaciers and the polar ice caps, so only about 1% of the world's water is available for human needs.

For centuries, humans have treated the earth's water supply as though it would last forever. They have wasted it by the trillions of gallons and have abused it for the sake of convenience, assuming that oceans, rivers, and lakes could absorb whatever waste was dumped into them. At the heart of this behavior is the belief that somehow water will dissolve or at least hide whatever is added to it, making poisons and garbage magically disappear. Only in the 20th century have people really begun to appreciate the fact that the water supply is neither bottomless nor immune to abuse.

In 1969, the Cuyahoga River in Cleveland—filled with oil, chemicals, and floating garbage—caught fire, and the five-story-high flames destroyed two bridges. Billowing foam caused by the widespread use of nonbiodegradable laundry and cleaning detergents appeared on streams and rivers, and photos of dead fish washing up onshore reinforced the public's perception that the nation's waters were far from pure. In 1988, a million gallons of diesel fuel spilled into the Monongahela River, contaminating the drinking water of almost a million people living in the Pittsburgh, Pennsylvania, area.

People soon found that protecting clean water was not an easy task. Water pollution is largely the result of the industrialized life-style, and controlling it requires changes in that life-style. To make things even more difficult, the sources of pollution are incredibly diverse and not always identifiable. Point sources—specific locations where contaminants are released, such as the discharge pipe from a sewage treatment plant or petroleum refinery—can be uncovered, but cleaning up or even simply reducing their output is a difficult and expensive process. Nonpoint sources—where contaminants cannot be traced to a single point, such as rainwater washing over croplands and carrying pesticide and fertilizer residues into nearby streams—offer even more complex problems. And when rivers, lakes, or pollutants cross international boundaries, who is ultimately responsible and who pays the cost of cleanup?

THE TIP OF THE ICEBERG

While some threats have been controlled and rivers no longer look like bubble baths, water continues to be polluted and misused, and the problems have become more difficult to solve. Newspapers now feature new but equally graphic images of pollution, including photos of the oil-coated victims of the Exxon *Valdez* spill that dumped 11 million gallons of oil into Alaska's pristine Prince William Sound on March 24, 1989, and of the medical waste and raw sewage that soiled East Coast beaches during the summer of 1988.

Underground storage tanks offer another example of the magnitude of the water pollution problem. The U.S. government estimates that there are at least 1 million underground petroleum and gasoline storage tanks in the United States. Based on a study by the Environmental Protection Agency (EPA), 20% of these may be leaking. That is 200,000 leaking underground storage tanks, or LUSTs. The gasoline and chemicals leaking from these tanks into the ground can contaminate wells and the groundwater supply. Congress has set up the Leaking Underground Storage Tank Trust

This stagnant pool in Mexico City may be the only source of water for poor people in the area. Undoubtedly, it is a breeding ground for the germs that cause typhoid and dysentery.

Fund to clean up leaks and contaminated soil and water, but the process will take years and probably billions of dollars.

Threats to the water supply and aquatic habitats are not an exclusively American problem but are found throughout the world, in both industrialized and developing nations. According to Earthscan, a news service of the United Nations, as many as 18,000 lakes in Sweden were affected by acid rain by 1982. Earthscan also reported that the Norwegian fish population had been declining in many lakes and rivers. Repeated exposure to acid rain kills fish, aquatic plants, and microorganisms and reduces the ability of salmon and trout to reproduce. This destroys aquatic habitats and severely affects local fishing industries. More than 70% of the pollution causing the acid rain in Scandinavia comes from Great Britain and western Europe, making it difficult for Sweden and Norway to control the problem.

In Eastern Europe, where nations are struggling to forge new economies, the challenge of preserving clean water is even greater. In Poland, half of the water supply is considered too polluted even for industrial use because the pollutants, primarily salt from mine runoff, would corrode pipes. In Czechoslovakia, 70% of the nation's rivers are heavily polluted and almost a third have no fish. Some of these eastern European rivers empty into the Baltic Sea, sharing their poisons with all the countries along the shoreline.

In the Soviet Union, Siberia's 12,200-square-mile Lake Baikal has been under assault from pulp-and-paper factories that have created a polluted zone 23 miles wide by discharging their wastes directly into the lake. At 5,712 feet deep, the lake is the world's deepest, largest by volume, and perhaps its oldest as well. It holds 80% of the Soviet Union's fresh water and 20% of the world's supply. Three-fourths of Lake Baikal's 2,500 fish and plant species exist nowhere else, including the Baikal nerpa, the world's only freshwater seal. In 1987, after 20 years of protest by fishermen and local residents, President Mikhail Gorbachev finally announced plans to replace one of the paper-producing plants with a nonpolluting furniture factory and to make the remaining plant safer for the environment.

Mismanaged water supplies are also spreading disease throughout developing countries in Africa, Asia, and Latin America. The creation of reservoirs and irrigation canals that trap standing water for use in agriculture, combined with the widespread lack of sewage treatment or even plumbing in many developing countries, has created favorable conditions for the spread of disease. Schistosomiasis, a group of tropical diseases of the blood caused by parasitic worms, is estimated to afflict 200 million people in 70 countries. Other waterborne diseases include amoebic and bacillary dysentery, hepatitis, yellow fever, cholera, typhoid fever, and malaria. More than a billion people in developing countries are affected by diseases spread by polluted water. In fact, three-quarters of all human disease is related to organisms spread by untreated water.

A QUESTION OF QUANTITY

The water issue is not only one of pollution but also of scarcity. On the one hand, the supply of clean water is diminished by pollutants. On the other hand, usable sources are being drained through wasteful consumption. Americans squander the most water. The average American uses two to four times as much water as does the average European, and Americans pay less for water than do the citizens of any other industrialized country.

Once the population of an area grows beyond the point where local water resources can support it, the community must look elsewhere to find more water to sustain its continued growth. Rivers must be dammed to create reservoirs, often depriving communities downstream of what they see as their fair share, or wells must dip deeper into the groundwater aquifers below.

Half of Americans depend on aquifers—naturally formed underground reservoirs—for their drinking water. However, in 35 states, aquifers are being tapped faster than they can be naturally replenished by rainwater seeping down through the soil. These so-called overdrafts can cause underground reservoirs to shrink permanently as the land above subsides or sinks to fill the empty space. Fresh water may also become contaminated in aquifers located near coastal areas when salt water seeps in to fill the void.

The Ogallala aquifer, the underground reservoir beneath parts of Colorado, Kansas, Nebraska, New Mexico, Oklahoma, and Texas, supplies millions of gallons of water to the arid West. However, the demands of expanding industry, intensive agriculture, and the influx of people to the Southwest threaten to empty this important water source. The Southwest has only 6% of the nation's naturally renewable water supply, but the region accounts for almost a third of the nation's demand. Planners have looked as far north as the Great Lakes for a potential source of water to supply the region's growing needs.

A rapid population increase has forced the city of Istanbul, in Turkey, to dump untreated garbage and sewage into its harbor.

One of the biggest problems facing water conservation is the lack of foresight shown by those deciding how to use or "manage" water. The fate of the Aral Sea in the Soviet Union is a lesson in what can happen. In the 1950s, Soviet planners decided to create an enormous cotton-growing region, partially diverting the Syr Darya and Amu Darya rivers for irrigation. Although the rerouting of these rivers opened up new cropland, it also caused the Aral Sea, into which these rivers normally feed, to shrink in area by 40%. Soviet planners knew this would happen but felt the additional cropland was more important. What they did not anticipate were the salt flats the receding salt water left behind. As a result, the surrounding land has been poisoned by storms that blow salt over the soil, diminishing the amount of cotton that can be produced.

AN INTERLOCKING WEB

Water is part of a larger interlocking web of environmental problems. For example, the acid rain that poisons lakes, rivers, and streams is created primarily by burning fossil fuels in automobiles and industrial operations. By reducing the use of fossil fuels, switching to alternative energy sources and cleaner fuels, installing pollution controls in smokestacks, reducing automobile use, and requiring more miles per gallon, the nitrogen oxides and sulfur dioxide that make rain acidic will be reduced.

Problems with waste disposal also have a significant impact on water supplies. Toxic chemicals and other pollutants leach out of landfills as rainwater filters down through the garbage. These contaminants pollute groundwater aquifers or surface water in nearby streams, rivers, lakes, and oceans. By building landfill sites with an impermeable liner made of clay and synthetic materials and by installing pipes to pump liquids out of landfills for treatment and disposal elsewhere, water supplies can be protected.

Paper recycling protects water supplies in two ways: It reduces deforestation, and it minimizes the amount of dioxins discharged by paper mills in the papermaking process. Cutting down trees to make so-called virgin or new paper increases the amount of soil that erodes and washes into streams, rivers, and reservoirs, clogging waterways and harming fish and other aquatic life. Deforestation also increases the severity of droughts and floods. Forest trees and plants absorb and hold rain in the soil and help control the runoff from storms that can cause flooding.

In addition, using chlorine to bleach wood pulp to make virgin paper creates a deadly chemical called dioxin, which is dumped into streams and rivers, causing cancer in fish and birds that live downstream. Manufacturing recycled paper results in significantly less dioxin pollution because less bleaching is required. In addition, new bleaching techniques are being used by European paper mills, replacing chlorine with oxygen, peroxide, or sodium hydroxide.

Global warming will probably also have a significant effect on water supplies, although the total impact is difficult to predict because scientists disagree about how much or even if temperatures will rise. As the climate heats up in the United States, water levels will most likely drop in certain areas as evaporation increases. Some wildlife habitat will be lost, and less water will be available to dilute pollutants, causing water sources to become polluted more quickly. In addition, if droughts increase in the midwestern Farm Belt, demands for alternative water sources will increase as the local aquifers are drained.

In addition, global warming and the earth's oceans are closely linked. Oceans absorb an enormous amount of carbon dioxide (CO_2), the most important greenhouse gas. Carbon dioxide is absorbed through the biological processes of phytoplankton, reducing the severity of global warming. In fact, some scientists believe the impact of these tiny creatures on the earth's climate is even greater than that of the trees of the earth's rainforests, which also store a great deal of carbon. These single-cell plants—the largest is a few thousandths of an inch wide—live in the surface layers of the ocean as far down as sunlight penetrates and serve as primary producers in the aquatic food web. Other multiple-cell plankton, such as krill, graze on phytoplankton. These krill are in turn eaten by whales and other ocean dwellers.

Through photosynthesis, the phytoplankton use the sun's energy to convert CO_2 and water into carbohydrates, fats, and proteins. After phytoplankton die, they sink to the bottom of the ocean, taking their carbon with them. This removes the CO_2 from the atmosphere for hundreds of years. Ocean currents slowly bring carbon back to the surface, where it returns to the atmosphere through evaporation. Some scientists believe that of the 6 billion tons of carbon released into the atmosphere by human activity each year, 3 billion tons are taken up by phytoplankton and removed to the ocean depths.

Much is still unknown about phytoplankton and their role in climate control. For example, will phytoplankton be able to absorb an ever-increasing amount of carbon dioxide as the human population continues to grow? What is the effect of oil spills and other toxic wastes dumped into the ocean on phytoplankton and their ability to reproduce and photosynthesize? And how will the increased ultraviolet light reaching the ocean's surface through a depleted ozone layer affect the phytoplankton population? These questions can be answered only as the effects of global warming and other environmental hazards become more apparent.

Water is a renewable resource. Normally, it is replenished through natural processes. However, if water is used faster than it can be replenished naturally, it will be depleted both in quantity and quality and could eventually be changed into a non-renewable resource. Fresh water was once abundant throughout much of the world, but it will become increasingly scarce in coming years. Rapid population growth will mean greater consumption, depleting the total water supply. The world's population is projected to reach 6 billion by the year 2000, up from 1.6 billion a century earlier.

The answer to the world's water crisis depends on how people perceive the problem. Taking a human-centered view, most people have assumed the world's resources exist exclusively to serve human needs. However, more and more people are realizing that the only way to deal with all of our environmental problems is to consider the welfare of the planet as a whole instead of simply from a human perspective. The so-called sustainable-earth view challenges the assumption that people have an inborn right to nature's resources and instead emphasizes respect for nature and the recognition of the rights of other species to exist. If people live more simply, resources will not be depleted as quickly, the amount of waste and pollution will be reduced, and habitats and endangered species can be protected.

These tall-sailed traditional feluccas identify this river as the Nile. The Aswân High Dam has disrupted annual flooding patterns and forced Egypt into dependence on chemical fertilizers to maintain its agricultural productivity.

chapter 2

WATER   RESOURCES

The 1% of the world's water supply that is fresh water
and available for human use is found underground, in rivers and
lakes, in the soil as soil moisture, in the atmosphere as vapor, and
in living things. This supply of fresh water is continually recycled
through the *hydrologic cycle*, or water cycle. The hydrologic
cycle has been recycling the same fixed supply of water throughout history, purifying and redistributing it in the process. The
cycle consists of three stages: evaporation, precipitation, and
runoff. When the sun warms the earth, water evaporates from the
land and oceans and forms vapor that rises into the atmosphere.
Through evaporation, water rids itself of salt, pollutants, and other
substances. This water vapor is transported great distances
through air currents in the atmosphere until it condenses and falls
back to earth as precipitation—rain, snow, hail, or sleet.

Once back on land, water either is soaked up by the
ground or flows off the surface as runoff, replenishing rivers and
lakes; much of it eventually flows back to the oceans. The water
in the ground either is stored as soil moisture, evaporates from the
soil surface, or percolates downward to the water table, where it
is stored in groundwater aquifers. Aquifers are not empty

underground caverns, but large areas of porous rock that hold water like a sponge. Water is also taken up and *transpired* by plants, which transfer water vapor to the atmosphere from their leaves and exposed parts. Then the process repeats itself.

The different stages of the water cycle occur at different rates, depending in part on where the water is located. On average, water in rivers is replaced every 18 to 20 days through the process of evaporation, precipitation, and runoff. Atmospheric moisture is replaced more quickly, about every 12 days. Lakes, depending on their size and depth, can take anywhere from 10 to 100 years to renew themselves, and renewal of deep underground aquifers requires several hundred years or more. While the shallow upper layers of the ocean are replaced every 100 to 150 years, the deepest layers can take 30,000 to 40,000 years. Replacement times are important because they reflect how long different water resources take to purify themselves or recover from environmental stress.

AQUATIC ECOSYSTEMS

Aquatic ecosystems include oceans, ponds, lakes, streams, rivers, coral reefs, estuaries, and coastal and inland wetlands. These ecosystems differ in the amount of salts dissolved in the water, in the depth of sunlight penetration, and in average water temperatures. Each of these ecosystems supports an enormous community of organisms—plants, animals, fish, and other creatures.

Oceans eventually receive most of the water that flows from rivers. Because of their size and strong currents, oceans can mix and dilute many pollutants, making them less harmful or even harmless. Oceans are divided into two major zones: coastal areas and open sea. The coastal zone extends from the high-tide mark on land to the edge of the continental shelf, an underwater shelflike extension of the continental landmass. Coastal zones represent less than 10% of the total ocean area, but they contain 90% of all marine plant and animal life as well as some of the earth's estuaries and coastal wetlands.

Estuaries are areas where fresh water and salty seawater mix, usually at the mouths of rivers and streams. Coastal wetlands, such as mangrove swamps and salt marshes, are places where the land is flooded all or part of the year with salt water, usually extending inland from estuaries. Estuaries and coastal wetlands are among the world's most productive ecosystems. They supply food and serve as spawning grounds for many species of fish and shellfish. They also serve as breeding grounds and habitats for birds and other wildlife.

Coastal areas also dilute and filter out waterborne chemicals, helping to cut down on pollutants that would disrupt swimming, fishing, and wildlife habitats. In addition, estuaries and coastal wetlands help protect entire coastal areas from the impact of the ocean by absorbing waves caused by violent storms and floodwaters from overflowing rivers.

FRESHWATER RESOURCES

Major freshwater resources include lakes and reservoirs, streams and rivers, inland wetlands, and groundwater aquifers. Lakes are natural bodies of standing fresh water formed by precipitation, water runoff from land, and water flowing from aquifers. Reservoirs, on the other hand, are artificial bodies of water, often formed behind dams to collect river water for crop irrigation or for drinking.

Rain that has not soaked into the ground or evaporated remains on the land as surface water. As this water flows over the land, it becomes runoff and flows into streams and rivers, eventually flowing downhill to the oceans to complete the hydrologic cycle. A *watershed*, or drainage basin, is the entire land area that channels the water via streams to a major river and ultimately to the sea. When the water runs over this watershed, it picks up sediment and other dissolved substances, including pollutants from the land. The relationship between land and water is important because it affects water quality. Runoff picks up nutrients from the land that help support plant life in rivers, lakes, and estuaries, which in turn support animal life. When runoff picks up toxic chemicals, on the other hand, aquatic plant and animal life can be destroyed.

Inland wetlands have special functions similar to coastal wetlands, although in this case they are lands covered with fresh water and located away from coastal areas. Inland wetlands include bogs, marshes, swamps, mud flats, and so-called prairie potholes. Inland wetlands provide habitats for a variety of fish, waterfowl, and other wildlife. Most waterfowl harvested in North America breed in the wetlands of Canada. Wetlands near rivers help control flooding and drought by storing water during periods of heavy rainfall and releasing it slowly. They also protect water quality by absorbing and diluting pollutants, and they are suitable for growing crops such as cranberries, blueberries, and rice.

Groundwater is a vital source of fresh water, especially in dry climates, where surface water is undependable or non-existent. Groundwater is stored in aquifers that are replenished as

In various areas of the United States, beautiful pristine lakes such as this one are losing their capacity to sustain aquatic life because of acid rain.

precipitation slowly seeps downward through soil and rock. The top surface of an underground reservoir is called the water table. During periods of drought, people say that the water table is getting low, because groundwater is being taken from the aquifer but is not being replaced quickly enough.

WATER USE

Water use is measured by *withdrawal* and *consumption*. Water is withdrawn when it is taken from a groundwater or surface-water source and transferred to a home, factory, farm, or other point of use. Water is consumed when it is not returned to its original source and becomes unavailable for reuse. For example, when a toilet is flushed, the water may be piped through a sewer, purified at a treatment plant, and then emptied back into the river it was withdrawn from. This water was withdrawn but has not been consumed. However, when house plants are watered, that water eventually enters the atmosphere when it is transpired by the leaves. This water is withdrawn and consumed because it is not returned to its original source.

Irrigation consumes the highest percentage of withdrawn water (approximately 55%), because so much evaporates or is transpired by the crops. The withdrawal/consumption equation can vary greatly within a country, depending on the distribution of agriculture and industry. For example, the eastern and western regions of the United States have significantly different impacts on their local water resources, although withdrawals are almost equal. In 1980, 52% of water withdrawn from local resources in the West was consumed; 48% was returned to its source after use. At the same time in the East, only 12% was consumed; 88% was returned. This is because 88% of water withdrawals in the West went to agriculture, whereas only 6% in the East was used for irrigation of crops.

People use water for many purposes in the home and office, in industry, in agriculture, and for recreation. In the United States, almost 80% of withdrawn water is used to irrigate farmlands and cool electric power plants. Domestic and municipal use includes water for drinking, food preparation, sanitation, cleaning, watering gardens, and various service industries, such as laundries, restaurants, and medical facilities. These uses account for about 7% of total water withdrawals. Though the quantity required for these purposes is not large, water quality must be excellent.

Domestic use also varies according to living standards. People in developing countries may use very little water, often five gallons or less per person each day, because water is scarce and the systems for delivering safe water to these rapidly growing populations are not well developed. People in industrialized nations, on the other hand, may use up to 125 gallons or more per person each day. Domestic and municipal water use is highest in the United States; domestic water use in Canada and Switzerland is also high.

Industry uses water in large amounts for cooling, processing, cleaning, and removing industrial wastes. Different manufacturing processes use significantly different amounts of water, depending on the technology used as well as the climate. As one would expect, use is higher in warm climates. This water is generally returned to the water cycle, but it is often heavily polluted. Currently, the United States and the Soviet Union account for half the world's industrial water use. Japan, Germany, and China follow.

Agricultural irrigation is the highest overall user of water. According to the World Resources Institute, water withdrawals for irrigation account for 82% of total water withdrawals in Asia, 41% in the United States, and 30% in Europe.

In industrialized countries, water is usually delivered to homes through the municipal water system or through wells that tap groundwater aquifers. In earlier centuries, towns and cities often took water directly from local rivers, which also received wastewater after use. When local sources became too polluted, cities reached out and tapped remote, pure sources, often in the mountains. They built dams or siphoned water off rivers and allowed gravity to bring the water to the city via water pipes or channels. Depending on the water quality, this water may be filtered, further disinfected with chlorine, and then delivered to local homes after a dash of fluoride has been added to curb tooth decay.

In the rural areas of industrialized countries, water is often delivered by private or municipal wells that tap into the groundwater. If the water is under pressure in the aquifer, that pressure will often be enough to push the water to the surface. If the pressure is not sufficient, pumps can be used.

Running water cannot be taken for granted by all American citizens. In fact, at least 4 million Americans lack access to clean water. In some small towns and rural regions, especially in the Appalachians, municipal water systems do not reach the rural population. Wells cannot tap into much groundwater because rain rolls swiftly off the granite mountains and does not sink into aquifers. Water is obtained from streams, where it is often contaminated, or from rainwater flowing off roofs along with leaves and bird droppings, or from water tank trucks. Obtaining a dependable supply of clean, fresh water in these areas is a constant challenge.

Obtaining clean water is an even greater challenge in most developing countries. According to the World Health Organization (WHO), in 1983 approximately 61% of people living in rural areas and 26% of urban dwellers in developing nations did not have access to safe drinking water. The WHO estimated that at least 5 million people die every year from preventable waterborne diseases such as cholera, dysentery, and diarrhea. In many such countries, poor people in drought-ridden areas spend many hours each day trying to find water, often taking it from contaminated streams and rivers. Urban dwellers must often pay high prices to buy water of undependable quality from water tankers at high prices. If a poor urban neighborhood is served by any sort of municipal water system, this often means one faucet for the entire community. And this water may also be contaminated. The United Nations Economic Commission for Latin America and the Caribbean estimates that less than 2% of total urban sewage in Latin America receives treatment. This untreated sewage is dumped into local rivers, which are used as sources of water.

Water is not distributed evenly around the world. Whereas some areas have a plentiful supply of water, arid and semiarid countries are constantly undersupplied and are vulnerable to droughts. In developing countries, rapid population growth and poor land use intensify the effects of droughts. In order to get enough food and fuelwood, poor people strip the land of trees, let their livestock overgraze grasslands, and grow crops on mountainsides at higher, more erosion prone elevations. These activities increase the severity of droughts because when the land is stripped of vegetation it cannot absorb as much rain.

Future water shortages could have a significant impact on the stability of many arid regions. Seventy percent of the world's 214 major river systems are shared by at least 2 countries, and 12 are shared by 5 or more countries. These countries often clash over water rights. Water scarcity in the arid Middle East is a continuing source of conflict. In 1967, Israel went to war in part because Arab countries were trying to divert water from the Jordan River. In 1990, Turkey threatened to cut off the Tigris and Euphrates rivers, which provide a significant portion of Iraq's water supply, unless Iraq withdrew from Kuwait.

Many experts consider the continued availability of fresh water one of the most serious long-term problems facing many parts of the world, including the United States. Water shortages and droughts in the western regions of the United States have become an ever-increasing threat in recent years as these areas grow rapidly.

Hoover Dam at Lake Mead, Arizona, on the Colorado River. Dams provide hydroelectric power but disrupt the natural flow of rivers, causing water shortages in some communities and prolonged legal battles over water rights.

chapter 3

THREATS TO THE WATER SUPPLY

The world's water supply is faced with two separate but related problems: pollution and scarcity. Three main groups of users affect both its quality and quantity: agriculture, industry, and municipalities. Each creates different threats to the water supply and requires different actions to reduce its negative impact. Agricultural irrigation uses and wastes by far the most water, often overusing local sources with disastrous results. When irrigation water is pumped from underground aquifers faster than it is naturally replenished, the ground above the aquifer can sink or subside. In California's San Joaquin Valley, 50 years of over-using local groundwater has caused an area the size of Connecticut to subside by as much as 30 feet in some places. Not only do farmers need to practice water conservation; they also need to protect the local supply from pollution. Agriculture-related pollution includes pesticides, herbicides, and fertilizers that run off the land into local streams and groundwater. Excessive irrigation also washes salts from the soil into local streams and rivers, making water too salty for users farther downstream.

Industry may waste less water than does agriculture, but the water it puts back into local water sources is sometimes severely polluted with hazardous substances. Toxic waste disposed of in landfills can seep into surface water or groundwater and poison the drinking supply. Water recycling and waste reduction can make significant improvements, but many companies are reluctant to make the financial investment.

Municipal water users must learn to practice water conservation techniques and become more sensitive to the pollution caused by household products and activities. For example, pouring one quart of used motor oil down a storm sewer can poison many gallons of water, and chemical fertilizers and pesticides spread on lawns and gardens can run off into local lakes and streams or seep down into the groundwater.

Other threats to the water supply and aquatic ecosystems include the filling in of wetlands for agricultural land or real estate development, overfishing, and the dumping of plastics into the ocean. Deforestation, in addition to destroying wildlife habitat and promoting global warming, increases the amount of silt that erodes from the cleared land, burying aquatic spawning grounds and clogging waterways so rivers no longer run freely.

THE EFFECTS OF WATER POLLUTION

Water pollution has many ill effects, both on aquatic ecosystems as a whole and on the health of humans and wildlife. Different types of water pollution affect the environment in different ways. Sewage or fertilizers that run off into lakes cause *eutrophication*, an explosive growth of algae and microorganisms that uses up the oxygen in the surrounding water, killing fish, aquatic mammals, and eventually the lake itself. On the other hand, most areas can recover from oil spills with time. The oil is buried or washed away, and populations of wildlife rebound. However, if an area is exposed to recurring spills, eventually the ecosystem will become permanently damaged. Scientists are still uncertain about the long-term effects of oil spills on marine life.

In the industrialized world, the most serious threat to people is water contaminated with hazardous chemicals. When the local water supply becomes contaminated, researchers often find increased rates of cancer and miscarriages. In developing countries, the problem is more often water contaminated with untreated sewage. Drinking such impure water causes many devastating, often fatal diseases. It has been estimated that three-quarters of all human disease is related to waterborne disease organisms, which kill 25 million people each year, most of them young children.

Wildlife is also in jeopardy. The North Sea, which receives pollution from industries along the Rhine River, used to have a large population of seals. Toxic chemicals in the water have disrupted the seals' ability to reproduce and weakened their defenses against disease. In 1988, about 16,000 seals succumbed to a virus. Now scientists have been able to count only about 15 to 20 seals. While many toxic chemicals have been banned, they continue to persist in the water. The beluga whales that live at the mouth of the St. Lawrence River in Canada have been devastated by pollution in that waterway. Twenty-five toxic chemicals have been discovered in these whales.

A very complicated water-related issue is hydroelectric power. Hydropower is trumpeted by many as a safe, cheap, nonpolluting form of power that will reduce our dependence on fossil fuel. In a hydropower plant, water flows through a set of turbines, turning their blades. The turbines are connected to generators, which rotate and produce electricity. The water is then returned to the river. To control the flow of water, dams are built across rivers. Besides channeling water for hydropower, dams also control overflow during flood season, provide water for irrigation, and create lakes that serve as recreation areas for fishing, boating, swimming, hiking, and camping.

Conservationists, however, often view hydropower projects with horror. In order to build the dam and create the reservoir that hydropower requires, large areas of land must be flooded, sometimes causing the relocation of thousands of people and the destruction of wildlife habitat. The flow of the entire river is disturbed, which may cause flooding upstream and diminished flow downstream. Fish migration to spawning grounds may be interrupted. Fish ladders and other devices are sometimes built to help deal with this problem, but they are only partially successful. Dams have been blamed in part for the 90% drop in the salmon population in the northwestern United States. Many conservationists feel that the environmental impact of new dam projects is overlooked or ignored in the quest for cheap power.

Major hydroelectric projects have been built or are being planned in the developing world as well. The Sardar Sarovar Dam in Gujurat, India, is part of a development scheme that would include the construction of 30 major dams, 135 medium-size dams, and more than 3,000 small dams over the next 40 to 50 years to provide irrigation water and hydroelectric power. Sardar Sarovar would displace more than 1.5 million people, mostly of minority origin.

Egypt's Aswân High Dam was built on the Nile River in the 1960s to provide flood control and irrigation water for the lower Nile basin and electricity for Cairo and other parts of Egypt. Ninety-five percent of Egypt is desert, and this water was badly needed. In some ways, the dam has been a success, supplying about one-third of the country's electricity. Year-round irrigation

Dead fish in Albufera, a lagoon near Valencia, Spain, are the result of agricultural chemicals applied to nearby rice fields.

has increased crop production, allowing farmers to harvest crops three times a year on land previously harvested only once a year. One million acres of desert have come under cultivation.

However, the dam has had numerous undesirable effects. The river used to flood land in the Nile basin every year. This yearly flooding fertilized the basin with silt and minerals and swept away snails that cause schistosomiasis. Now that the dam has stopped the flooding, the cropland must be treated with commercial fertilizer. The new fertilizer plants use up much of the electrical power produced by the dam. Salts that were once washed away have built up in the soil. This salinization has offset three-quarters of the gain in food production, because soil affected with excessive salts produces much less food. In addition, the Nile used to discharge a lot of sediment where the river met the ocean. Because the river now discharges so little sediment, the ocean is eroding the delta and taking with it many acres of agricultural land. This nutrient-rich silt used to feed great schools of sardines, mackerel, shrimp, and lobster. These food sources have all but disappeared. Finally, the area around the dam suffered a fairly severe earthquake in 1981. Scientists think that since this is a low-risk area, the quake was triggered by the weight of the water in the reservoir, Lake Nasser.

Clearly, the benefits of dams and reservoirs must be weighed against their costs. They do provide a clean and dependable form of energy. However, the reservoirs behind dams fill up with silt and become useless in 20 to 200 years, depending on local conditions. The flooding of land behind the dam displaces people and destroys vast areas of valuable agricultural land. Finally, faulty construction, earthquakes, and sabotage can cause dams to fail, causing floods and devastating destruction.

The issues surrounding hydroelectric power projects are very complex. People often compete with other living things that depend on the same water sources. In the case of California's Mono Lake, the water needs of Los Angeles—275 miles to the south—are at odds with the hundreds of thousands of birds that depend on the lake's brine shrimp to fuel their long flight to winter feeding grounds in South America.

In the 1950s, in order to serve a rapidly growing population, the Los Angeles Department of Water and Power tapped into four of the five freshwater streams that feed into Mono Lake. As a result, the lake's water level fell 40 feet. The lake's total area has shrunk from 86 to 61 square miles, exposing 16,000 acres of lake bottom. As the water level dropped, the salinity of the lake sharply increased. Today the water is two and a half times as salty as the ocean. Many aquatic species have died.

Brine shrimp still provide food for thousands of birds. As many as 50,000 California gulls—95% of those in the state and 20% of all the gulls in the world—come to the lake every year to nest and feed on the shrimp. Millions of other birds, more than 100 species, use Mono Lake as a stopover point as they migrate between breeding and wintering grounds. Scientists have counted 800,000 birds feeding at Mono Lake in a single day. However, if the lake becomes increasingly saline, will the shrimp population eventually collapse? And if the shrimp fail, where will the migrating flocks go?

Two-thirds of California's water supply is in the northern part of the state, but Los Angeles sprawls across the arid south and thus is forced to look north for water. The streams that used to

Slash-and-burn forest clearing in the Amazon. The rainforest is a natural sponge, and once it is gone, heavy tropical rains degrade the soil and cause floods downstream.

feed Mono Lake deliver 32 billion gallons of water each year through the Los Angeles Aqueduct, supplying 17% of the city's water and 2% of its electricity. Replacement sources are not easy to find.

Beginning with the winter of 1981, the years were wet enough so that the runoff from precipitation in the feeder streams exceeded the capacity of the Los Angeles Aqueduct. The excess water then started to raise the level of the lake, reversing its decline. Some trout accompanied the renewed flow in Rush

Creek, the largest of the feeder streams. Researchers estimated that by 1984, 30,000 brown and rainbow trout inhabited this section of Rush Creek. But dry years will return, and continued diversion of Rush Creek to the Los Angeles Aqueduct will once again dry up the river and kill most of the fish. California law prohibits any activity that would endanger an existing fishery. This law implies that continued diversion of the water during dry years would be illegal.

The Audubon Society brought the matter to court, claiming that the destruction of the fishery and ecological damage to Mono Lake violated the public trust and that withdrawal of water should be stopped. The Los Angeles Department of Water and Power argued that the people of Los Angeles needed the water. The court was unable to accept either argument and left the conflict unresolved, ruling that "some responsible body ought to reconsider the allocation of the waters of Mono Basin." The issue has remained unsolved in spite of continued legal action. The difficulty lies in the traditional view that water is a public asset that must be put to the best use. Is the best use providing water to the millions of people who live in Los Angeles or preserving Mono Basin, an important aquatic habitat?

*At Kimberly, Idaho, a modern low-pressure sprinkler system helps con-
serve irrigation water while avoiding soil degradation.*

chapter 4

WATER   AND
AGRICULTURE

Agriculture is the largest single consumer of water. Worldwide, agricultural irrigation accounts for about 70% of total water use. Between 1950 and 1985, the total land area under irrigation worldwide nearly tripled. The Water Resources Council expects it to increase even more by the year 2000. In the United States, agriculture consumes around 89 billion gallons of water every day. Most of that water is used for the irrigated farms concentrated in the 17 western states.

As such an enormous consumer of water, agriculture poses problems to both water quantity and quality. More than 50% of the water used for irrigation does not return to its source, but is consumed by evaporation, seepage, or runoff. Excessive irrigation can degrade fields by washing nutrients out of the soil, eventually reducing the ability of the land to support growth. In addition, runoff from croplands and barnyards, carrying chemical pesticides and fertilizers as well as animal waste, is a significant nonpoint source of pollution that contaminates groundwater and affects streams, ponds, and lakes.

Agriculture also has a devastating impact on wetland habitats. To replace degraded cropland or to increase the amount of land under cultivation, wetlands are dredged or filled in. According to the Environmental Protection Agency, agriculture was responsible for 87% of wetland losses from the mid-1950s to the mid-1970s. As explained earlier, these forested wetlands, inland marshes, and wet meadows serve important and often overlooked roles. They help improve water quality by absorbing pollutants. They reduce flood and storm damage by absorbing excessive precipitation during heavy storms and later releasing it gradually. They provide important spawning grounds for fish and nesting areas for waterfowl.

The Florida Everglades, for example, is a unique and fragile ecosystem. Its natural cycle includes a complex system of rapid growth during the wet season, followed by excessive dryness and fire, and finally drought. The plants and animals of the Everglades have adapted to this cycle, and the existence of all these species is interconnected. If one element of the ecosystem is affected adversely, the whole swamp will feel the impact. This delicate cycle is threatened by local agriculture. To the north of the Everglades, dairy farms, sugarcane fields, and orange groves lie in former glades that were drained for cultivation. When farmers fertilize their crops, the chemicals run off into local streams and eventually enter the marsh. In addition, sewage from the cities flows into the Everglades. All this organic waste promotes rapid growth. But if the grasses and other plants grow too rapidly, the delicate relationships among plants and animals may be disturbed. Pesticides also spill into the Everglades, killing plankton, fish, reptiles, and birds.

In addition, large quantities of water are pumped from Lake Okeechobee, the northern water source for the Everglades, for irrigation and domestic use in the fields and cities of south Florida. Some of this water never returns to the marshes. All these factors could alter the plant-animal balance and change the Everglades forever.

Attempts have been made to protect remaining wetlands. The Farm Act of 1985 includes a provision that withholds agricultural subsidies from landowners who convert wetlands to croplands. The latest farm bill, signed in late 1990, is designed to enroll 40 million to 45 million acres in a wetland reserve. However, much wetland area is still threatened.

AGRICULTURAL RUNOFF

Agriculture is a major source of pollutants from fertilizers, pesticides, herbicides, and animal wastes. These are generally termed nonpoint sources because they enter the water through runoff from the land over large areas and therefore the exact source of pollution cannot be pinpointed or easily controlled. In the United States, nonpoint pollution from agriculture is responsible for an estimated 64% of the total mass of pollutants entering rivers and 57% of those entering lakes.

Pesticides and fertilizers have contaminated the surface water and groundwater in many areas of the United States, as well as in the Canadian food-growing regions. Groundwater contamination from agricultural chemicals is a matter of immediate concern to farm families and other rural residents who rely on these sources for drinking water. An EPA survey in 1984 found that two-thirds of rural household wells tested violated at least one federal health standard for drinking water. The most common contaminants were nitrates from fertilizers.

This problem extends beyond the western agricultural regions. Long Island, New York, has experienced groundwater pollution by a pesticide called Temik. In the 1960s and 1970s, local farmers found Temik very effective against a variety of pests in local potato fields. However, because much of the soil on Long Island is sandy and porous, rainwater carried Temik into the drinking water supply. In 1982, 17,000 wells in eastern Long Island were tested, and 2,300 were found to exceed the EPA's

In California's famous Imperial Valley, one of the richest growing regions in the country, salt-laden irrigation water forms a small river as it drains from the fields.

health advisory level for Temik. Temik has also been found in the groundwater of Florida, California, and Missouri.

When agricultural chemicals enter rivers and lakes, their effects can become even more toxic because of the process of *biomagnification,* or the increase in the concentration of toxic chemicals as one moves up the food chain. The original concentration of a pesticide in the water may be quite low. However, microorganisms store and concentrate the pesticide in their tissues. Fish eat the microorganisms and further concentrate the poison with each daily intake. The higher the organism is in the food chain, the greater the concentration of poison. Eating the fish swimming in these waters could make birds, animals, and people sick.

ORGANIC WASTES AND EUTROPHICATION

In the United States, livestock are the principal source of the organic wastes that pollute surface water and groundwater, producing about five times as much such waste as do humans. This runoff carries potassium, phosphates, and nitrogen compounds into water supplies, resulting in excessive nitrogen concentrations leading to eutrophication.

Eutrophication occurs when these organic nutrients stimulate the growth of algae and microorganisms. The algae and microorganisms consume the organic pollutants, initially helping to purify the water. However, the algae also consume large quantities of dissolved oxygen. As the oxygen levels in the water drop, fish and other aquatic creatures perish. Eutrophication is reversible, but if the sources of organic pollutants are not checked, recovery may be impossible. Many lakes and rivers throughout the world are polluted in this manner. In 1958, Lake Erie, more than 10,000 square miles in area, was reported to be dying because of eutrophication. Through pollution control efforts, the lake now shows signs of recuperating, but pollutants still pour in. Fertilizer runoff is difficult to curb.

EFFECTS OF IRRIGATION

Traditional forms of intensive irrigation require an enormous amount of water, waste a lot of it, and have adverse effects on the soil as well as water supplies. In the gravity flow systems used on most of the world's irrigated cropland—where gravity pulls water through channels between rows of crops— about half of the water used is consumed by seepage and evaporation without ever being taken up by the crops.

Besides wasting water, overwatering leads to salinization and waterlogging, two devastating side effects of irrigation. When poorly drained fields are irrigated excessively, the water table—the top surface of the groundwater aquifer—rises. If the water table rises too near to the surface, the water fills the air spaces in the soil and suffocates the plant roots. The soil also becomes more difficult to cultivate. As water evaporates, it leaves behind salts and minerals. The accumulation of salts and minerals on the soil's surface is called salinization. Most plants cannot grow in salty soil, so the fertility of the land decreases. Salinization may also make the soil hard and impenetrable.

One-tenth of all irrigated land worldwide suffers from waterlogging. Productivity on this land has fallen by about 20%. Approximately 25% of the world's total irrigated acreage is

These furrow dams trap and hold rainwater, allowing it to seep into the soil at a more controlled rate, thus conserving water and stopping soil erosion.

suffering a loss of productivity because of increased salinization. In the United States, salinization has reduced productivity on 25% to 35% of all irrigated land in the 17 western states. In India, one-third to one-half of all irrigated land is threatened by salinization and waterlogging. Worldwide, an estimated 50% to 65% of all currently irrigated cropland is projected to suffer reduced productivity by the year 2000. Also, as irrigation water is repeatedly withdrawn from and returned to a river, the salinity of the river increases. Irrigating with this water only accelerates the rate of salinization.

Worldwide, 63% of the water used for irrigation is wasted; only 37% contributes to the growth of crops. Improving irrigation techniques would not only reduce water consumption but would also reduce the adverse effects of intensive agriculture. In the United States, an enormous amount of water is wasted for irrigation in the western states because the cost of water is heavily subsidized by the government. The U.S. Bureau of Reclamation sells water at very low rates. This policy was designed many years ago to provide low-cost irrigation water to small family-owned farms in order to develop the arid West. Over time, however, farms in many western states became agribusinesses, large-scale farming corporations. Simply pricing water realistically could boost irrigation efficiency significantly.

Improved irrigation efficiency would help conserve water supplies and prevent or reduce damage to croplands. Lining irrigation ditches with cement saves a lot of water and money by reducing seepage, as does the use of irrigation pipes. Farmers can also use microirrigation techniques, such as drip or trickle systems that deliver water through perforated tubes directly to crop roots. These systems are three times more effective than traditional methods of irrigation. Computer systems can help monitor soil moisture content. The use of such methods increased nearly eightfold between 1974 and 1982, but they are still used on less than 1% of all irrigated cropland.

Less expensive improvements can also be made, including the leveling of cropland to reduce runoff and the recycling of excess water using pumps and ponds. In addition, by releasing irrigation water at intervals instead of continuously, water and energy use can be reduced by as much as 40%.

OVERMINING AQUIFERS

Overmining groundwater aquifers to irrigate crops and support large populations in the arid West is a serious problem. Overmining or overdrafting means that more water is withdrawn from underground reservoirs than can be naturally replaced. Groundwater has a very slow refill or recharge rate. As mentioned earlier, the Ogallala aquifer is seriously threatened. Water withdrawn from the Ogallala aquifer irrigates one-fifth of all U.S. cropland. This aquifer has an extremely low natural recharge rate because the region it underlies has such low precipitation. Currently, the overall rate of withdrawal is eight times its natural recharge rate. Even higher withdrawal rates are taking place in certain parts of the aquifer beneath Texas, New Mexico, Colorado, and Oklahoma. Experts project that, at the present rate of withdrawal, much of the Ogallala aquifer will be empty by the year 2020. The problem is not exclusively American. Nations such as Jordan and Israel also use groundwater aquifers to supply a large amount of their water needs. In these areas of little rain, renewal is very slow and overdrafting is a continuing problem.

One result of groundwater overmining is subsidence. Groundwater fills pores in the soil and thus supports the soil above the aquifer. When water is withdrawn faster than it is replaced, the soil compacts and subsides, or sinks. Subsidence has damaged pipelines, railroads, highways, homes, factories, and canals. In Florida and other southern states, groundwater depletion has created huge sinkholes that may measure up to 330 feet across and 165 feet deep.

THE COLORADO RIVER

Agricultural water use not only takes a toll on groundwater reservoirs but can also have a severe impact on surface water. From its headwaters high in the Rockies, the Colorado River supplies water to more than 17 million people in Wyoming, Utah, Colorado, Nevada, New Mexico, Arizona, California, and Mexico. As it flows from Colorado to Mexico, large quantities of water are consumed, and the river picks up a heavy burden of sediment and dissolved salts from agricultural runoff. By the time the Colorado reaches the Gulf of California, it is little more than a salty trickle. At its headwaters in the Rockies, the salt concentration is 40 parts per million. When it reaches Mexico, the concentration is over 800 parts per million.

The diversion and degradation of the Colorado has led to a long-standing dispute between the United States and Mexico over the use of the river. In order to improve the quality of the water that reaches Mexico, the Yuma Desalting Plant was constructed by the U.S. government in Arizona to remove some of the salt contaminating the water. However, the diminished quantity of water flowing to Mexico is still a serious problem.

The federally financed $3.9 billion Central Arizona Project began pumping water from the Colorado to Phoenix in 1985. It began delivering water to Tucson in 1991. Arizona has been legally entitled to one-fifth of the Colorado River's annual flow since 1922. Before the Central Arizona Project, the state was unable to get more than half its share. Although diverting water from the Colorado will partially replace groundwater overdrafts in many parts of Arizona, the added claim on the Colorado has caused Southern California to lose up to one-fifth of the water that it had diverted from the river.

Southern California has explored various options for fulfilling its water needs. One proposal, rejected in 1982, was to expand the state's water system by building a canal to divert much of the water that flows into San Francisco Bay to Southern California. Opponents of the $1 billion plan said it would degrade the Sacramento River, threaten fishing, and reduce the natural flushing action that cleans San Francisco Bay of pollutants. They also pointed to the amount of water already wasted by the south, claiming that a 10% increase in irrigation efficiency would provide enough water to fulfill the domestic and industrial demands of Southern California.

Much of the frustration expressed by conservationists stems from the type of agriculture practiced in Southern California. Pastureland gets the most water. The production of grass and hay for cows and sheep consumes the same amount of water as do all of the state's 27 million inhabitants. Alfalfa and other water-intensive crops, such as rice and cotton, require as much water as the San Francisco and Los Angeles metropolitan areas combined. By switching to less water-intensive crops and using water-saving irrigation techniques, water use could be cut considerably, with the savings passed on to the growing urban population. So far, however, farmers and ranchers have little incentive to conserve. As water shortages continue to affect California cities and towns, however, pressure may increase to change California water policy.

REDUCING AGRICULTURE'S IMPACT

Trading water rights is one way the American West is trying to overcome its water shortages. This approach often involves urban areas making deals with farmers or land companies with excess water supplies to sell. Water trading has some drawbacks. For example, rural areas can export too much of their stream flows and end up ecologically degraded. However, trading water rights allows farms to switch from water-intensive crops, such as alfalfa, without losing money because they can sell the water they conserve. In another deal with multiple benefits, the Metropolitan Water District of Southern California agreed to finance rural water conservation projects, such as fixing leaks in irrigation canals. In exchange, the agency receives the conserved water.

Improving overall agricultural efficiency and reducing its environmental impact will take more than a few water deals. Agricultural inputs—water, fertilizers, pesticides, and machinery—must be used more efficiently. Improving irrigation techniques and drainage systems reduces water consumption as well as the negative effects of waterlogging and salinization. Using crops that are well suited to local conditions—instead of trying to grow pastureland in the desert—will also help reduce dependence on excessive irrigation.

Decreasing our dependence on chemical pesticides and fertilizers will help reduce pollution from agricultural runoff. Sustainable farming methods emphasize using organic fertilizers,

In Fresno, California, a system of underground plastic tubes developed by soil physicist Claude J. Phene supplies water directly to the plant roots, conserving water and producing record tomato harvests.

such as animal and crop wastes, to improve soil fertility and water retention. Chemical pesticides can be replaced with a system of integrated pest management (IPM). By bringing in the natural predator of a pest, the pest population is controlled without chemicals. Another strategy of IPM is to sterilize and release male pests in their natural breeding grounds so that the females cannot produce the next generation of insects and the pest population dies off. By growing different species of plants together instead of planting one enormous field with a single crop, natural diseases and pests cannot spread as quickly or destructively.

Unfortunately, mechanized farming and federal water subsidies provide little incentive to change agricultural practices. However, as the public becomes more concerned about the existence of pesticides in drinking water and on produce, and as competition increases among farmers, cities, and industry for limited water resources, changes will be necessary.

This waste sludge from a paper mill has been treated with alum so that the sludge will separate from the water before it is returned to the river. Papermaking can create highly toxic waste products, but new technologies are helping to minimize pollution.

W A T E R   A N D   I N D U S T R Y

Worldwide, industrial water use accounts for about 21% of total water withdrawal. Up to 80% of this water is used for cooling, especially during the production of energy. Other industrial uses include processing, cleaning, and removing industrial wastes. The industrialized regions of North America, Europe, and the Soviet Union make the greatest demands in this area, but water use by industry in the developing countries of Asia, Africa, and Latin America is projected to increase dramatically by the end of the 20th century.

Many industrial processes produce waste, some of which is toxic even in the smallest quantities. Major polluters include the chemical, food-processing, petrochemical, pulp-and-paper, and textile industries. These industries poison water either through the direct release of waste into nearby rivers and oceans or through the creation of waste dumps, from which toxic chemicals leach into groundwater. These wastes are usually heavy metals or synthetic organic compounds, neither of which can be easily absorbed by the surrounding environment.

In developing countries, industrial waste is generally uncontrolled, and in some areas water quality has been severely degraded. For example, in the Maipo river basin in Chile, only a small percentage of the wastewater produced by local chemical, plastic, and rubber industries is treated before it is discharged into the river, and none of the waste from the pulp-and-paper industry is treated at all. As controls against pollution get stricter in the developed nations, companies look to the developing world for places to relocate their industrial plants. Environmental regulations in developing countries are less demanding. Other companies look to the developing world as a convenient dumping ground for their hazardous waste.

However, even though environmental regulations are stricter in industrialized countries, the threat of industrial pollution in these countries is very real. Disaster can strike even with the best intentions. In November 1986, a fire broke out in the warehouse of the Sandoz Corporation, a chemical company in Basel, Switzerland, along the banks of the Rhine River. Firefighters doused the fire with thousands of gallons of water, which mixed with the pesticides, herbicides, and other chemicals in the factory. As a result, 30 tons of pesticides spilled into the Rhine in the form of a poisonous purple sludge, which killed virtually all the aquatic life in the river as far as 120 miles downstream. Many towns that drew their drinking water from the river had to shut down their treatment facilities.

In order to avoid the expensive treatment process required to detoxify chemical wastes, some companies have been tempted to seal material in drums and dump the drums at sea. However, the drums eventually corrode and leak, poisoning marine life. The Dutch government reported that the North Sea is among the most polluted in the world. High levels of toxic industrial chemicals known as PCBs, or polychlorinated

An old textile mill on the Passaic River, near Paterson, New Jersey. Factories are often situated along waterways for the convenience of dumping their industrial wastes.

biphenyls, have caused massive fish poisonings and high rates of infertility, miscarriages, and death among local seal populations. Another method of disposing of some kinds of hazardous waste is to burn them at sea. Though the process destroys much of the waste, some remains intact and ends up in the ocean. The North Sea is currently the only place where this kind of ocean incineration takes place.

OIL POLLUTION

The petroleum industry affects water quality in many ways, although oil spills are the most evident. The American public was outraged in March 1989 when the Exxon *Valdez* spilled 11 million gallons of crude oil into Prince William Sound in Alaska. The effects of that spill on the local environment will not be known for years, in part because the results of scientific studies have been sealed, pending court cases against Exxon.

Many different factors affect the impact of an oil spill and the speed with which the environment recovers, including ocean currents, water temperature, and the strength of tides. In the case of the Amoco *Cadiz* spill, which dumped 68 million gallons of oil off the Atlantic coast of France in March 1978, scientists found that most of the major effects of the spill had disappeared within three years.

Researchers have found certain general patterns of recovery. On exposed rocky beaches with a lot of wave action, such as the ones on France's Atlantic coast, little oil is left after a year. The oil persists for two to three years on quieter beaches, where it often mixes with sand and is buried. Salt marshes are the most severely affected, and cleanup efforts are more destructive than helpful. Fish and bird populations usually rebound, but there may be long-term effects on ecosystems, and these effects are as yet poorly understood. Of course, estimates of recovery times assume that the environment is recovering from one oil spill. If an area is subject to frequent spills, such as the Red Sea or the Persian Gulf with their busy tanker traffic and military activity, ecosystems might be damaged permanently.

In response to the Exxon *Valdez* disaster, legislation was finally passed during the summer of 1990 that strengthened existing federal laws dealing with oil spill prevention, liability, and compensation. Under the new bill, shipowner liability increased from $150 per gross ton of tanker capacity to $1,200 per gross ton. The bill also broadens coverage for damages to include the "loss of use" of natural resources. It creates a $1 billion trust fund for cleanup and compensation and requires double hulls on all oil transport vessels. Most existing tankers must comply within the next 20 years. The law also puts the federal government in charge of cleanup and removal of all major spills instead of the responsible company, although the company is liable for the cleanup costs. Furthermore, the bill calls for the creation of Coast Guard spill-response teams in 10 separate districts. One of the many problems with the Exxon *Valdez* spill was the delay in getting cleanup teams and equipment to the site to begin work.

In general, the frequency of oil spills has decreased dramatically since the 1970s. However, major spills are not the only way oil finds its way into the ocean. In fact, oil spills from tanker accidents account for less than one-third of all the oil released. Most oil pollution results from washing tankers out with seawater and releasing the oily residue into the ocean. In many parts of the world such activity is illegal, but it is hard to monitor. Other spills result from accidents at offshore drilling sites, routine losses during handling at seaports, leaks from pipes, and even the dumping of used oil into storm drains and city sewers.

BIOREMEDIATION

One of the most innovative solutions for dealing with oil spills and other kinds of hazardous waste is *bioremediation*, or the use of bacteria to digest toxic substances. These bacteria either exist naturally or are genetically engineered. The technology has been around for decades, but its use is becoming more widespread as scientists understand it better.

The idea behind bioremediation is to speed up a natural process. Sewage treatment plants use bacteria to consume organic wastes, boosting the oxygen content of the polluted water to stimulate bacterial growth and speed up the cleanup process. Wherever the naturally occurring population of waste-eating bacteria is too small to handle a serious toxic spill, bioremediation engineers create a bacterial population explosion by adding oxygen, nitrogen, phosphorous, or whatever elements are essential to bacterial growth.

Once the microbes have consumed the offending chemicals, artificial fertilization is stopped and the bacterial population returns to normal. Biotech firms mass-produce

Foam caused by the dumping of potash salts in the Alsace region between France and Germany pollutes the Rhine River all the way to the North Sea.

different strains of bacteria to be used for specific cleanup tasks. Although this process may sound like the perfect solution, it works only under certain conditions. Bioremediation is also often slower than conventional cleanup methods, so it may not be appropriate when fast action is needed. However, one clear advantage of bioremediation is price; it is significantly cheaper than conventional methods.

Bioremediation was tested on some of the beaches of Prince William Sound in the wake of the Exxon *Valdez* spill. Preliminary results have been encouraging. Exxon researchers sprayed beaches with a fertilizer that stimulated the growth of naturally occurring bacteria that consume the hydrocarbons in oil. Within two weeks, the sprayed beaches showed dramatic improvement compared with untreated areas. However, some drawbacks remain. Biological activity declined sharply with the return of cold weather. Scientists still need to determine how effective the technique was in digesting oil that seeped below the surface of pebble beaches or penetrated porous stone.

THE MISSISSIPPI

In the April 16, 1990, issue of *Newsweek,* a special team of correspondents reported on the state of the Mississippi River. The Mississippi runs for 2,500 miles from upper Minnesota to the Louisiana delta, carrying 100 trillion gallons of water every year, and is therefore America's major waterway. For this very reason, it has become a center of industrial activity. Many power plants have situated themselves near its banks, drawing river water to cool their generating systems. And where there are power plants along a river, there are bound to be barges carrying coal and oil, with all the risks of spillage. At Genoa, Wisconsin, is the La Crosse Boiling Water Reactor, a nuclear power plant that has not produced any electricity since 1987 but which is kept open only so that river water can be used to cool radioactive waste buried under the plant. There is at the moment no other place to store this waste.

At Sauget, Illinois, just below St. Louis, a dozen chemical and pharmaceutical plants discharge their waste into the Mississippi in such quantities that the regional office of the EPA labeled the water the most toxic they had ever tested. The Monsanto Corporation is working hard to reduce production of pollutants such as benzene and paradichlorobenzene, by-products in the manufacture of toilet bowl fresheners and mothballs. In the area around Clarksdale, Mississippi, north of Greenville, the river picks up pesticide and herbicide residues from the surrounding cotton fields.

The area of the river between Baton Rouge and New Orleans in Louisiana is known as the Chemical Corridor, with 25% of the nation's chemical industry located there. Concessions made by the state government to attract this industry have been called by Governor Buddy Roemer "a deal with the Devil." There are significantly higher rates of death from cancer and other diseases among the people who live in this region.

REDUCING WATER POLLUTION

Industry does have some avenues for pollution control. If wastewater is discharged to municipal treatment plants, factories must pretreat the water to remove toxic pollutants that municipal treatment facilities cannot handle. Pretreatment removes heavy metals and toxic chemicals, including such by-products of industrial processes as lead from the manufacture of batteries. The electroplating industry can remove heavy metals from wastewater and reuse the gold, silver, copper, zinc, and cadmium that are normally washed away during processing. Reducing waste of this kind lowers costs. Factories can also install systems to reuse this water instead of returning it to local streams and rivers. Industry also needs to reduce the amount of water it uses as well as the amount of waste it discharges.

There are some success stories. Pioneer Metal Finishing, an electroplating company in New Jersey, cut water use by 96% and sludge production by 20%. Drastic water treatment efforts and the introduction of water recycling in factories along the Rhine have lead to significant decreases in heavy metals in the river. The 3M Corporation instituted a company-wide pollution prevention effort in the mid 1970s. During the 12-year effort, waste generation was halved, for a total savings of $300 million. In Sweden, which is a leader in industrial wastewater treatment, a pharmaceuticals company called Astra improved in-plant water and materials recycling and substituted water for solvents, cutting toxic wastes by half. And Cleo Wrap, a wrapping-paper company, replaced solvent-based ink with water-based ink, virtually eliminating hazardous waste.

THE VALDEZ PRINCIPLES

After the Exxon *Valdez* spill, the Coalition for Environmentally Responsible Economies (CERES), released the Valdez Principles on September 7, 1989, as guidelines for improving corporate responsibility for ecological damage.

Corporations can agree to comply with the principles, thereby making a public commitment to protect the environment. The Valdez principles are as follows:

1. *Protection of the Biosphere.* Minimize the release of pollutants that may cause environmental damage.
2. *Sustainable Use of Natural Resources.* Conserve nonrenewable natural resources through efficient use and careful planning.
3. *Reduction and Disposal of Waste.* Minimize the creation of waste, especially hazardous waste, and dispose of such materials in a safe, responsible manner.
4. *Wise Use of Energy.* Make every effort to use environmentally safe and sustainable energy sources to meet operating requirements.

Oil sludge covers this underwater photographer who has surfaced from a dive near a wrecked tanker.

5. *Risk Reduction.* Diminish environmental, health, and safety risks to employees and surrounding communities.

6. *Marketing of Safe Products and Services.* Sell products that minimize adverse environmental impact and that are safe for consumers.

7. *Damage Compensation.* Accept responsibility for any harm the company causes to the environment; conduct bioremediation, and compensate affected parties.

8. *Disclosure.* Public dissemination of information relating to operations that harm the environment or pose health or safety hazards.

9. *Environmental Directors and Managers.* Appoint at least one board member who is qualified to represent environmental interests.

10. *Assessment and Annual Audit.* Produce and publicize each year a self-evaluation of progress toward implementing the principles and meeting all applicable laws and regulations worldwide. Environmental audits will also be produced annually and distributed to the public.

*A ship operated by the Thames Water Authority in England dumps
sewage sludge into the North Sea.*

chapter 6

MUNICIPAL WATER USE

Cities, towns, and private homes account for about 8% of total water use. However, both municipal and domestic use have a significant negative effect on the water supply. Municipal pollution comes from both point and nonpoint sources. The three major point sources are sewage treatment plants, storm drains, and power plants.

Sewage treatment plants take raw sewage from homes, offices, factories, and sewers and try to neutralize chemical and bacterial contamination. They remove most but not all of the pollutants and dump the wastewater into local rivers, lakes, or oceans. When heavy rains overcome the capacity of sewage treatment plants, the overflow of raw sewage is often dumped directly into local water supplies without treatment. Boston Harbor, one of the most polluted harbors in the United States, has 86 sewer discharge pipes along the shoreline. About 40 to 50 times a year, runoff from storms overwhelms the treatment system, and raw sewage pours directly into the harbor from these pipes. Also, as sewer systems age, cracks may develop in the pipes, allowing raw sewage to seep into the ground.

Power plants produce a form of pollution—thermal pollution—that is often overlooked because it is hard to see or smell. Electric power is produced by generating heat to produce steam, which pushes the blades of electric turbines. Water is used in large quantities to cool and recondense the steam, and in the process, large amounts of heat are transferred to the outside water source. Power plants take cool water from a nearby river or lake, pass it through the plant, and return the heated water to its original source. The water may not be contaminated with pollutants, but it is heated, and this can disrupt aquatic ecosystems. Large rivers that flow rapidly usually suffer little damage, but when a single plant or a number of plants dump large amounts of heated water into the same lake or a slow-moving river, fish and other organisms can go into shock and die.

Warmer temperatures decrease the amount of dissolved oxygen in the water and also cause aquatic organisms to increase their rates of respiration. This makes them more susceptible to disease, parasites, and toxic chemicals. When heated water is discharged into shallow water near shore, it may disrupt spawning grounds and kill young fish. Thermal water pollution can be reduced by limiting the number of power plants discharging heated water into the same body of water. Heated water must be kept away from the shore zone. Special cooling towers or shallow cooling ponds and canals can be employed before water is returned to lakes and rivers.

Nonpoint sources of water pollution are also important to identify, although they are more difficult to control. One threat to groundwater is leachate from municipal landfills. As rainwater drips down through landfills, it dissolves chemicals in the solid waste, which then flow into groundwater aquifers or local rivers and streams. Another significant source of nonpoint pollution is urban and suburban runoff. Precipitation, such as rain or melting snow, carries many pollutants from roads, parking lots, lawns, and construction sites into the local water supply. Pesticides and fertilizers on lawns and gardens, salt and deicing compounds on roads, puddles of oil and gas, and even heavy metals are all part of urban runoff. Anything that is spilled on the ground may eventually make its way into the water supply.

OCEAN POLLUTION

The ocean ultimately receives a large share of the world's polluted waters. Not only is it the recipient of pollution from rivers and streams; it is also affected by increasing urban development along the world's coastlines. Greater coastal development means that fragile and productive coastal ecosystems will come under ever-increasing assault. In the United States, sewage treatment plants are one of the most prevalent point sources of coastal pollution. About 35% of all U.S. sewage ends up in the ocean. The National Oceanic and Atmospheric Administration (NOAA) reported that the oceans received 3.3 trillion gallons of sewage from sewage treatment plants in 1980. The NOAA estimates that this volume will increase to 5.4 trillion gallons by the year 2000.

In addition to waste from sewage treatment plants, the oceans receive agricultural and urban runoff and garbage and untreated sewage from ships and barges. They also absorb industrial wastes, sludge dredged or scraped from the bottoms of rivers and harbors to maintain shipping channels, accidental oil

Both municipal sewage and industrial waste flow into the Passaic River from this "point source" of pollution in Newark, New Jersey.

spills, and intentional discharges of oil from tankers emptying their bilges.

Another threat to marine life comes from plastics dumped into the ocean. Discarded plastics are responsible for the death of up to 2 million seabirds and 100,000 marine mammals each year. Sea turtles consume plastic bags that they mistake for jellyfish and choke to death. Birds mistake plastic pellets floating on the surface for fish eggs or larvae. Seals become entangled in fishing nets.

In deep water areas, oceans can dilute, disperse, and break down large amounts of sewage, sludge, and some industrial waste. Even so, marine mammals and fish show the deadly effects of ocean pollution. Coastal areas bear the brunt of ocean dumping, and they are much more fragile than the open sea. Many American bays and harbors are badly polluted, including New York Harbor, Chesapeake Bay, Puget Sound, San Francisco Bay, the Gulf of Mexico, and Santa Monica Bay. Fifty-six percent of the coastal areas surrounding Great Britain are seriously polluted, as are much of the Mediterranean and Baltic seas.

CHESAPEAKE BAY

Chesapeake Bay is the largest estuary in the United States and one of the world's most productive. It is home to about 200 species of fish and shellfish. The bay lies in the heart of one of the major commercial and industrial regions of the United States. It has a vast drainage basin that includes 9 large rivers and 141 smaller rivers in parts of 6 eastern states. Between 1950 and 1985, the population near the drainage basin increased by 50% as urban communities expanded.

The estuary receives wastes from point and nonpoint sources throughout this area, such as sewage treatment plants, urban and suburban runoff, agricultural runoff, and toxic waste from industry. In many areas of the bay, underwater grasses that helped to control erosion and filter pollutants have virtually disappeared. Populations of oysters, crabs, and some commercially important fish have fallen sharply. Severe eutrophication and oxygen depletion exists in many areas.

Through the cooperative efforts of citizens, state and federal officials, and industries, things are changing in Chesapeake Bay. Since 1983, more than $650 million in federal and state funds have been spent to clean up the bay. Discharges of phosphates from point sources, a key ingredient in oxygen depletion, have dropped about 20%. Maryland, Virginia, Pennsylvania, and the District of Columbia have tightened regulations on municipal treatment facilities. Maryland banned phosphate detergents in 1985, as did Virginia in 1987. However, laws are only as effective as their enforcement, so private citizens' action groups have initiated legal action to fight violations. A lawsuit was filed and won against Bethlehem Steel for illegally dumping wastes into the estuary. As encouraging as these actions are, the cleanup has barely begun. It will take decades to bring Chesapeake Bay back to life; costs could run between $1 billion and $3 billion.

WASTEWATER TREATMENT

In industrialized countries, most of the sewage and wastes from point sources are purified to some degree. In suburban and rural areas, sewage and wastewater from each house are often fed into a septic tank. The septic tank traps large solids and discharges the remaining wastes over a large drainage area. The soil filters out some potential pollutants as these wastes percolate downward, and bacteria in the soil decompose biodegradable materials.

In some small urban areas and in many developing countries, sewage is transported to wastewater lagoons, where solids settle out, and air, sunlight, and microorganisms break down wastes. Typically, water stays in the lagoon for 30 days and is then treated with chlorine and pumped out for use by a city or farm.

In large urban areas in industrialized countries, waterborne wastes flow through sewer pipes to wastewater plants. At the treatment plant, sewage goes through up to three levels of purification. Overall, the purpose of treatment plants is to remove solids—from rags and sticks to sand and smaller particles—as well as to reduce amounts of organic matter and pollutants and to restore oxygen so that the water returned to lakes and rivers can support life. Lakes and rivers naturally clean water in much the same way, but treatment plants are faster and can handle more waste. The three levels of sewage treatment are called primary, secondary, and advanced.

Primary treatment removes 40% to 50% of the solids. Screens filter out debris, such as sticks, stones, and rags, and a grit chamber slows down the flow of water and allows the sand and other heavy solids to settle at the bottom. Then a sedimentation tank allows smaller particles to settle to the bottom as sludge. Scrapers or other devices collect the sludge and any scum or grease floating on top of the tank.

Secondary treatment completes the process, so that 85% to 90% of the pollutants are removed. Combined primary and secondary treatment must be used in all communities in the United States served by wastewater treatment plants. Secondary sewage treatment uses bacteria to remove biodegradable organic wastes. An aeration tank supplies large amounts of oxygen to a mixture of wastewater, bacteria, and other microorganisms. The oxygen speeds the growth of microorganisms that consume harmful organic matter. A secondary sedimentation tank allows microorganisms and solid wastes to form clumps and settle. Before the wastewater leaves the treatment plant, a disinfectant, such as chlorine, is added to kill disease-causing organisms in the water.

Combined primary and secondary treatment still leave 3% of the suspended solids, 50% of the nitrogen, 70% of the phosphates, and 30% of most toxic metal compounds and synthetic organic chemicals in the wastewater. Virtually none of the pesticides are removed. In certain cases, advanced sewage treatment is used. This consists of a number of specialized chemical and physical processes that lower the quantity of the remaining pollutants. Types of treatment vary depending on the specific contaminants. Other purification techniques are being studied at different points in the sewage treatment process. For example, wastewater from secondary treatment can be purified further if forced to flow through canals filled with plants such as water hyacinths or bullrushes. These plants absorb toxic organic chemicals and metal compounds. However, disposal of the contaminated plants must still be dealt with.

Arcata, a city on California's Humboldt Bay, has an unusual, environmentally sound three-stage treatment system. About 10 years ago, Arcata was found to be actively polluting Humboldt Bay with the wastewater outflow from its conventional secondary sewage treatment plant. The town's solution was to create 94 acres of saltwater and freshwater wetlands.

After going through secondary sewage treatment, water is pumped into oxidation ponds that kill harmful bacteria through changes in acidity. During the day, plant photosynthesis consumes carbon dioxide in the water. During the night, the carbon dioxide is released. The consumption and release of carbon dioxide causes enough variation in acidity to kill bacteria. From the oxidation ponds, the water flows into the newly created marsh, where organisms in the wetlands filter the water naturally. Carefully selected plants further break down the wastewater and absorb heavy metals. In two or three weeks, the purified water is discharged into the bay.

Not only does Arcata's sewage treatment provide cleaner water than a traditional secondary treatment plant; it also enhances the appearance of the environment. The marsh, built on what was once the town dump, has attracted so many species of birds that the area has been declared a bird sanctuary by the Audubon Society. It has become an important and beautiful recreational area. However, Arcata's "natural" treatment is only natural in the tertiary phase. Local engineers are confident that their system would work without all the machinery used for secondary treatment, but the state of California will not let them try it out.

Using a different technology, the Center for the Protection and Restoration of Waters in Providence, Rhode Island, is experimenting with solar aquatics, a sun-based method of purifying waste without using hazardous chemicals. Solar aquatics imitates the way nature cleans water, only more quickly and thoroughly. Large, translucent, cylindrical tanks are placed in rows inside a greenhouse, connected by pipes that use gravity to pull the water through each line. Whereas the first tanks contain bacteria, algae, and snails, tanks downstream contain more complex life-forms, such as higher plants, fish, and other mollusks. In the first tanks, bacteria consume the organic material in wastewater. Algae thrive on nutrients released by the bacteria and grow rapidly. Snails then consume the algae. This cycle continues, with one organism consuming and being consumed, all the while cleaning the wastewater. In tanks farther on, plants float on the water, their oxygen-rich roots descending beneath the water's surface, where higher organisms graze. In the last tanks, fish swim around in the clean water, and the final purification takes place in a marsh.

Such low-tech sewage treatment techniques hold great promise for both industrialized and developing nations. They can reduce the energy consumption and chemicals required by conventional sewage treatment. Low-tech techniques also cost considerably less than sewage treatment plants. The high cost of sewage treatment is one reason it is so scarce in developing countries. Low-tech projects are therefore attractive. In Lima, Peru, for example, a pilot project is under way in a place called San Juan de Miraflores, making use of 21 settling and algae treatment ponds. Not only does this system give local farmers treated water to irrigate their crops; it also provides a new source of fish protein from the ponds.

THE GROWTH OF URBAN POPULATIONS

More and more people live in urban areas. In the 18th century, only about 3% of the world's population lived in cities. By 1950, 29% had moved to urban areas, as jobs shifted away from farming and rural areas. By 1985, more than 40% lived in cities, and by 2025 an estimated 60% of the world's population will be city dwellers. Most of the growth in urban populations is taking place in the cities of developing nations, which are least able to provide for the basic needs of this rapid influx of people.

One out of four urban dwellers in developing countries lacks easy access to uncontaminated water. In Lima, Peru, 2 million people get their water from tank trucks, an expensive delivery system. And even this water is not guaranteed to be pure. Others take their water from streams contaminated by urban runoff and untreated sewage. Despite recent improvements in

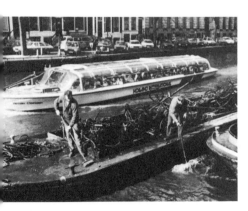

In Amsterdam, Holland, workers annually fish hundreds of bicycles from the canals in an effort to clean the popular waterways before the tourist season.

Central America, only 59% of the population have access to an adequate supply of clean water, and only 58% have access to waste disposal, such as latrines, septic tanks, communal toilets, or connections to public sewers.

Without adequate sewage disposal, a lot of sewage is left untreated, either dumped on land or into nearby rivers and streams that also provide drinking water. Coastal cities often dump untreated sewage into the ocean where people swim, wash, and fish. The lack of both sanitation and a safe water supply is a serious health problem and an issue of growing concern as these urban populations continue to increase.

Water quantity will continue to be a problem in both industrialized and developing countries as people concentrate in urban areas. Competition for water supplies will increase between municipal, industrial, and agricultural users. Water conservation will be become increasingly important. In Latin American cities, between 25% and 35% of the water consumed is lost through leaks and breaks in the water system. In New York City, residents have successfully fought the installation of water meters for a century. New Yorkers viewed their water as essentially free, and they paid for water using a formula that depended on building size or the number of fixtures rather than on actual use. Officials hope to install meters in every building in the city by the end of the 1990s.

Other improvements include requiring low-flow toilets in new buildings, setting lawn-watering and car-washing regulations, fixing leaks and cracks in water pipes and aqueduct systems, and recycling wastewater. Combined with water savings in industry and agriculture, water conservation efforts in cities could help supply the water needs of the next century.

In the Imperial Valley, California, a researcher studies melons grown on land previously irrigated with salt-laden wastewater.

G O V E R N M E N T   A C T I O N

Protecting the quantity and quality of the water supply offers many challenges. Water issues respect no state or international boundaries. However, progress has been made on many fronts in the last 20 years, as the number of water pollution and water conservation laws has increased, along with the seriousness of efforts to enforce those laws. In addition to federal laws, many states have drafted their own legislation in order to take control of local water quality. International cooperation in cleaning up and protecting waterways that cross borders has also increased.

In 1969, the United States Congress signed into law the National Environmental Policy Act (NEPA), which declared that the government was responsible for restoring and maintaining environmental quality. Prior to the signing of that bill, a number of laws aimed at protecting water quality had been passed, but the government had not adopted a strong environmental policy or an effective enforcement mechanism.

A number of important bills were passed in the 1970s. The Clean Water Act, passed in 1977 and reauthorized in 1987, was an amended version of the 1972 Federal Water Pollution

Control Act. The bill included provisions to control the pollutants discharged into waterways and provided grants to cities to help them improve their sewage treatment facilities to meet federal standards. As with all environmental legislation, however, the act has some loopholes. It does not require pollution control for storm sewers, which funnel overflow during heavy rains and carry sewage directly into city waterways.

Another important bill, the Ocean Dumping Ban Act, passed by the United States Congress in 1988, required that ocean dumping of industrial waste and sewage be phased out. Current dumpers must either end ocean dumping by 1991 or pay escalating permit fees as long as dumping continues. The 1972 National Coastal Zone Management Act, reauthorized in 1980, was designed to control the deterioration of coastlines brought about by development and population pressure. In response, state governments are working together to improve the management of coastal zones, with support from the National Oceanic and Atmospheric Administration. Twenty-eight of the 35 eligible states had developed federally approved plans by 1985.

Many important initiatives to protect the country's water supply and aquatic ecosystems are taking place at the regional and state level. Maryland, Pennsylvania, Virginia, and the District of Columbia are working together to reverse the decline of the Chesapeake Bay. New Jersey has proposed an ambitious plan to reduce ocean pollution. Washington State began a comprehensive marine pollution cleanup program in 1985.

The agency with the responsibility for enforcing most federal environmental laws is the Environmental Protection Agency. The EPA has broad powers, including the power to sue almost any U.S. citizen or company for violating antipollution laws and the power to charge fines of up to $25,000 a day for environmental violations. Unfortunately, enforcing environmental policy is not that simple. The job of managing the nation's environmental and resource policy is widely dispersed among many different agencies within the federal government's executive branch as well as among state agencies. This division of control limits the government's ability to develop an effective approach to interrelated problems.

One of the EPA's tools is the the Hazardous Waste Trust Fund, or Superfund. The Superfund is part of the Comprehensive Environmental Response, Compensation, and Liability Act of 1980 (CERCLA). The impetus behind CERCLA came from a number of dramatic incidents in the late 1970s that underlined the dangers of ignoring hazardous waste. The most notorious example was Love Canal. This community near Niagara Falls,

Agricultural technicians use an electromagnetic device to test soil salinity. Too much wastefully applied irrigation water tends to build up the salt content of soils.

New York, was evacuated after hazardous waste buried over a 25-year period contaminated groundwater and threatened the public health.

One of CERCLA's objectives was to develop a comprehensive program for cleaning up the worst hazardous waste sites. CERCLA was supposed to make responsible parties pay for cleanup whenever possible. However, for situations where responsible parties could not pay or could not be found, CERCLA set up the $1.6 billion Superfund to perform remedial cleanups and to respond to emergencies.

THE UNITED STATES AND ITS NEIGHBORS

Water has also been a matter of controversy among the United States, Mexico, and Canada. Mexico has protested for many years that U.S. agricultural interests claim too much water from the Colorado River, leaving only a trickle to cross the Mexican border. In addition, water quality is poor, with a high saline content and pesticide residues. To try to reduce tensions, the United States built a desalination plant in Arizona to remove some of the salt, but concerns over water quality and low volume continue.

Cooperation between the United States and Canada dates back to the 1909 Boundary Waters Treaty, established to address transboundary water issues. Chief among the two nations' concerns are the Great Lakes, which contain one-fifth of all the surface fresh water in the world and 95% of all the surface fresh water in the United States.

Geographer Stephen McRae and soil scientist James Rhoades study maps of soil salinity as part of a special project funded by the U.S. Department of Agriculture.

The Great Lakes ecosystem has been under attack for years. For two centuries, people have treated the lakes as a dumping ground. A 1985 joint report of the Royal Society of Canada and the U.S. National Research Council called the Great Lakes basin one of the most contaminated regions on the continent. Threats to the ecosystem include overfishing, the reshaping of shorelines and river mouths, the damming of rivers and the building of canals, pollution from shipping, and the draining of wetlands. In addition, pollution comes from a multitude of sources on both sides of the border, including fertilizer runoff, the discharge of industrial waste and hazardous chemicals, and sewage and household waste. Fish and birds in the area show a disturbing incidence of tumors and the inability to reproduce.

To help reduce the threats to the Great Lakes basin and begin restoring the aquatic habitat, the two nations signed the Great Lakes Water Quality Agreement in 1972. This first agreement emphasized the problems caused by phosphorus discharges in Lakes Erie and Ontario. A second agreement in 1978 continued to push for phosphorus controls but shifted the emphasis to controlling toxic chemicals. Amendments signed in 1987 broadened the agreement's scope to fight airborne toxic contaminants deposited in the lakes. Some progress has been made, particularly in Lake Erie, which is slowly recovering. However, a lot of work remains to be done.

Another area of concern for the two countries is the Gulf of Maine, stretching from Cape Cod to the Bay of Fundy. Threats to the gulf include overfishing, especially on Georges Bank, and the discharge of industrial effluents and wastewater. The Gulf of Maine Council on the Marine Environment is currently trying to set up a cooperative mechanism that would involve the U.S. and Canadian governments as well as each state bordering the gulf, with the objective of reducing coastal and marine pollution.

PROTECTING THE OCEANS

Protecting the marine environment is impossible without regional and international cooperation. So many countries border any one ocean that improvements in pollution controls by one country can be offset completely by another country's lack of concern. International cooperation is an enormous challenge and is affected by the political relations between countries as well as their economic health and environmental priorities. Even so, a number of international and regional programs have been initiated to control the flow of pollution into the oceans.

The Law of the Sea, MARPOL, and the London Dumping Convention are three of the most significant agreements limiting the disposal of wastes in international waters. The Third United Nations Conference on the Law of the Sea, first convened in 1973, completed its draft of the Law of the Sea Convention in 1982. With the participation of more than 150 nations, the convention's goal was to develop an all-encompassing treaty covering marine issues ranging from fishing and mining to navigation and pollution. Not surprisingly, the conference failed to agree on such a comprehensive treaty. However, the 1982 convention did address the issue of seabed mining and established Exclusive Economic Zones (EEZs) that recognized each nation's economic sovereignty over the oceans within 200 miles of its coast and obligated nations to protect and preserve their marine resources.

The 1973 International Convention for the Prevention of Pollution from Ships, known as MARPOL, established minimum distances from land for dumping sewage, garbage, and toxic waste. The convention also set strict limits on the amount of oil that ships can discharge at sea. An addendum to the agreement prohibits ocean dumping of plastics and requires ports to receive and handle trash from incoming ships. Unfortunately, this amendment lacked enough ratifying votes until 1987, when the United States finally cast a vote in its favor, allowing it to take effect.

The 1975 London Dumping Convention bans the dumping from all ships and aircraft of heavy metals, petroleum products, and known carcinogens. Other substances, including lead, cyanide, and pesticides, require special permits for dumping. Since 1983, the convention has also included a moratorium on dumping low-level radioactive wastes.

In 1974, the United Nations Environmental Programme (UNEP) has spearheaded the Regional Seas Programme, which was supported by 26 international organizations and 120 nations. Its first initiative was the highly successful Mediterranean Action Plan. Signed by 17 Mediterranean countries in 1976, the Mediterranean Action Plan tackled both seaborne and land-based pollution in the form of sewage, industrial waste, and agricultural pesticides. Since the beginning of the effort, Mediterranean waters and beaches have become cleaner and shellfish beds are less contaminated.

HELPING EASTERN EUROPE

One of the toughest challenges facing eastern European countries in the 1990s is rebuilding their shattered economies while at the same time cleaning up their ravaged environments. Many eastern European countries have virtually no pollution controls. Because the economic demands of rebuilding their countries are so great, these nations have been looking to the West for help in financing their environmental cleanup. They

Agricultural engineers in Oregon check the radio-controlled metering system of this advanced five-center pivot sprinkler system.

must build modern factories and control pollution at existing ones. They must rescue medieval buildings and statues being eroded by air pollution and provide medical assistance to the tens of thousands of eastern European children whose lungs have been scarred by air pollution.

At this point, western Europe has responded more quickly to these requests for aid than has the United States. The European Community plans to contribute $50 million to help clean up the environments of Poland and Hungary. Finland will contribute about $25 million more, and Sweden, Holland, and Germany are making similar contributions. So far, the United States has provided $15 million for a pollution control project in Poland and $5 million for a regional environmental center in Hungary. These contributions help, but EPA officials say Poland alone needs $5 billion to battle its environmental problems. Western Europe is not simply being altruistic in helping the former Eastern bloc countries control pollution. Pollution from the East causes acid rain in western European countries, especially in Scandinavia. Rivers polluted by eastern European industry flow into shared bodies of water, such as the Baltic Sea.

One lesson to learn from all these national, regional, and international efforts is that it would have been a lot cheaper to control pollution and conserve resources from the start. The question is, Will developing countries, as they become more industrialized, repeat the mistakes of the industrialized world, or will they be more environmentally conscious? Unfortunately, being environmentally sensitive takes more money at the outset in order to buy pollution control equipment and advanced sewage treatment plants; but it could save a lot of money in the end, not to mention reduced risks to public health.

A scientist checks soil and water samples to determine the concentra-tions of various chemicals.

chapter 8

PROTECTING THE FUTURE

Although water quality continues to be an important environmental concern, significant improvements have been made in pollution control and water treatment. New technology is being developed to detoxify hazardous wastes before they reach the dump and to clean them up once they seep into streams and groundwater. Farmers are starting to appreciate the need for more conservative farming techniques that cut down on pesticide and herbicide use and reduce water usage. Major environmental campaigns have targeted seriously polluted waterways with some success. The Cuyahoga River once again lures boaters and fishermen. The eutrophication in Lake Erie has been slowed. Fish have returned to the Cubatao River in Brazil after a 30-year absence. Yet the fight for clean water is not over. Without stronger laws and better enforcement, protecting our water supply will be difficult. As world population continues to increase, ensuring that people have enough water will become a pressing issue.

Besides protecting the water that exists, some countries are dreaming about new sources. One possible source for the future is glacial ice: towing a piece of a glacier to the place where it is needed and using the melted ice as a new water supply. So far, the price of such an operation has been prohibitively high.

Another potential residential water source is *gray water*. Gray water is the water that goes down the drain in the shower or sink. Some people collect gray water and use it on their lawns and gardens. A gray water reuse system requires the installation of a tank to collect the water from the pipes. In some states, gray water reuse is illegal. However, in times of drought, gray water reuse is often presented as a valuable conservation method.

Another, more expensive method for increasing water supplies, used in some arid parts of the world, is *desalination*. Desalination plants in the Middle East and North Africa produce about two-thirds of the world's desalinated water. Saudi Arabia has the most ambitious desalinated water program in the world, with 29 desalination plants. Desalination is also used in parts of Florida.

The two most widely used desalination methods are *distillation* and *reverse osmosis*. Distillation involves heating salt water until it evaporates and condenses as fresh water, leaving solid salt behind. Reverse osmosis uses high pressure to force salt water through a thin membrane that allows water molecules but not the dissolved salts to pass through its pores. The basic problem with all desalination methods is that they use large amounts of energy and are therefore expensive. Additional energy is required to pump desalinated water inland from coastal desalination plants. The most practical place for desalination is in the coastal cities of arid regions, where the cost of getting fresh water by any method is high. However, fresh water produced by desalination will probably never be cheap enough to irrigate large expanses of crops.

WATER CONSERVATION

The cheapest, most effective method for increasing the world's water supply is through water conservation. In areas where drought is a recurring problem, water conservation is a part of daily life, but it should become second nature to everyone. Farmers can install less wasteful irrigation systems that will also protect their land from waterlogging and salinization. Industry can cut down on water use and recycle what is used. And individuals can pay more attention to what they use at home as well. California provides an interesting—

Medical waste collected from the beaches of Staten Island, New York, in 1988. All the beaches in Staten Island and Brooklyn were closed when sewage poured into New York's harbor after a power failure shut down a local sewage treatment plant.

and ongoing—case study of the challenges of dealing with a diminished water supply.

In early 1991, California was facing its fifth straight year of drought. Reservoirs were dropping so low that they revealed old, abandoned towns and long-dead orchards flooded years ago when the state developed its vast water system. Forced by threats of water rationing, local companies, such as semiconductor manufacturer Intel, are experimenting with water recycling methods. Kelco, a San Diego–based chemical manufacturer, plans to cut water consumption by 40% over the next 3 years through recycling. Faced with drastic cutbacks in low-priced water from state and federally controlled reservoirs, some farmers are planting less-thirsty crops. California will survive the current drought, but will behavior change, or will people return to their wasteful habits once the rains come?

Cutting down on water use in the home is easy and quickly becomes a habit. On average, the bathroom claims 65% of residential water use (40% for toilet flushing), the laundry uses 15%, kitchen activities use 10%, and outdoor chores use about 10%. Here is a list of simple water conservation measures anyone can practice:

1. Turn off the water while brushing teeth or shaving. This saves one to two gallons each time.

2. Take short showers instead of baths. Whereas a bath uses 20 to 40 gallons of water, a 3-minute shower uses only 15 to 20 gallons.

3. Install low-flow showerheads and faucets. They can reduce water use by as much as 50%.

4. Install a toilet dam or place a half-gallon plastic bottle of water in toilet tank. This reduces the amount of water used for each flush.

5. Collect rainwater to water lawns and gardens. Water in the early morning or evening when it is cooler to avoid rapid evaporation. Do not overwater.

6. Fix leaky faucets and toilets. A small drip wastes 25 gallons a day.

7. Wash only full loads of laundry, not one pair of jeans at a time, and use the short cycle.

8. When washing dishes by hand, do not let the faucet run. Fill two dishpans and use one for washing and one for rinsing.

9. Keep a jug of water in the refrigerator instead of running tap water until it becomes cold enough to drink.

10. When running the faucet for hot water, catch the cool water in a pan and use it for cooking or to water plants.

Individuals can also have a significant impact on water quality. It must be remembered that everything that goes down the drain, the sewer, or even onto the ground eventually ends up in a river, the ocean, or someone's water supply. Here are some suggestions to help preserve water quality:

1. Never dump used motor oil down the sewer or into the trash. Find out about community hazardous waste pickup sites or oil recycling centers. This applies to other hazardous household chemicals, such as paint and paint remover, antifreeze, batteries, and lawn pesticides.

2. Avoid using liquid fertilizers, weed killers, and pesticides on lawn or garden. Find out about organic methods instead.

3. Use phosphate-free detergents. Phosphates cause algal blooms and eutrophication in ponds, lakes, and streams.

4. Boat owners should not throw trash overboard. Protect the shoreline by reducing speed and not producing wakes within 500 feet of the shore.

The Greenpeace ship Sirius *leads a fleet of Danish fishing boats in surrounding and harassing the chemical waste burn ship* Vesta *(the larger ship in the middle left background) as it attempts to burn and dump waste into the North Sea.*

5. Reduce household use of toxic cleaning products by substituting benign alternatives. What follows is a list of common household cleaning agents and possible substitutes:

Drain cleaner. To keep drains clear, pour in boiling water once a month. Put one-quarter cup baking soda and one-half cup cider vinegar down slow drains.

Household cleaners. Try baking soda and mild detergent, and use a little muscle.

Mothballs. Substitute cedar chips or newspapers; clean clothing before storing; wrap wool clothing in plastic bags during the summer.

Oven cleaner. Use salt or baking soda and water. For stubborn spots, let ammonia soak overnight.

Silver polish. Soak silver in a mixture of water, baking soda, salt, and small pieces of aluminum foil.

Toilet cleaner. Substitute mild detergent or small amounts of bleach.

Window cleaners. Wash with warm water and vinegar; dry with crumpled-up newspaper to avoid streaks.

The biggest challenge today is to keep environmental concerns alive and active. Constant vigilance is required if past wrongs are to be undone, new attitudes and behavior are to be encouraged, and future generations are to be assured the clean water necessary for survival.

APPENDIX: FOR MORE INFORMATION

Environmental Organizations

American Water Resources
  Association
5410 Grosvenor Lane, Suite 220
Bethesda, MD 20814
(301) 493-8600

Chesapeake Bay Foundation
162 Prince George Street
Annapolis, MD 21401
(301) 268-8816

Clean Water Action
  Project
317 Pennsylvania Avenue SE
Washington, DC 20003
(202) 547-1196

Clean Water Fund
National Office
1320 18th Street NW
Washington, DC 20036
(202) 457-0336

Friends of the Earth
Take Back the Coast! Campaign
218 D Street SE
Washington, DC 20003
(202) 544-2600

The Oceanography Society
1701 K Street NW
Washington, DC 20006
(202) 234-2109

Pollution Probe
12 Madison Avenue
Toronto, Ontario M5R 2S1
Canada
(416) 926-1907

Sierra Club Headquarters
"Gaining a Voice: Increasing
  Your Influence with Members
  of Congress"
730 Polk Street
San Francisco, CA 93109
(415) 776-2211

Soil and Water Conservation
  Society
7515 NE Ankeny Road
Ankeny, IA 50021

State Water Resources Control
  Board
Paul R. Bonderson Building
901 P Street
P.O. Box 100
Sacramento, CA 95812-0100
(916) 322-3132

Agriculture in the Classroom
Room 234-W
U.S. Department of
  Agriculture
Washington, DC 20250

Valdez Principles Coalition for
  Environmentally Responsible
  Economies
711 Atlantic Avenue, Fifth Floor
Boston, MA 02111
(617) 451-0927

Water Pollution Control
  Federation
601 Wythe Street
Alexandria, VA 22314-1994
(703) 684-2400

FURTHER READING

Canby, Thomas Y. "Our Most Precious Resource: Water." *National Geographic* (August 1980): 144–79.

CEIP Fund. *The Complete Guide to Environmental Careers.* Covelo, CA: Island Press, 1990.

Cobb, Charles E., Jr. "The Great Lakes' Troubled Waters." *National Geographic* (July 1987): 2–31.

Corson, Walter F., ed. "Fresh Water." In *The Global Ecology Handbook*, 155–71. Boston: Beacon Press, 1990.

Crawford, Mark. "Bacteria Effective in Alaska Cleanup." *Science* (March 30, 1990).

El-Ashry, Mohamed, and Diana C. Gibbons, eds. *Water and the Arid Lands of the Western United States.* New York: Cambridge University Press, 1988.

Fine, Christopher. "Freshwater." In *World Resources 1990–91*. New York: Oxford University Press, 1990.

———. "Oceans and Coasts." In *World Resources 1990–91*. New York: Oxford University Press, 1990.

———. *Oceans in Peril.* New York: Atheneum, 1987.

Hodgson, Bryan. "Alaska's Big Spill: Can the Wilderness Heal?" *National Geographic* (Jan. 1990): 5–43.

Keating, Michael. *To the Last Drop: Canada and the World's Water Crisis.* Toronto: Macmillan, 1986.

Kolbert, Elizabeth. "The Drink of Millions." *New York Times Magazine* (March 4, 1990): 30–34, 73, 93.

McPhee, John. *Encounters with the Archdruid.* New York: Farrar, Straus & Giroux, 1971.

Myers, Norman, ed. "The Oceans." In *Gaia: An Atlas of Planet Management,* 284–99. Garden City, NJ: Anchor Books, 1984.

———. "Superfund: Looking Back, Looking Forward." *EPA Journal* (Jan./Feb. 1987).

Pringle, Laurence. *Water—the Next Great Resource Battle.* New York: Macmillan, 1982.

Young, Gordon. "The Troubled Waters of Mono Lake." *National Geographic* (October 1981): 504–25.

GLOSSARY

acid rain Rain that has an abnormally high concentration of sulfuric and nitric acid; caused by industrial air pollution and automobile exhaust.

aquifer Porous underground rock or a rock formation that creates a natural underground reservoir holding **groundwater**.

biomagnification An increase in the concentration of DDT, PCBs, and other chemicals in successively higher levels of the food chain.

bioremediation The use of living organisms or microbes to clean up soil and water contaminated by toxic waste accidents.

desalinize To remove dissolved salts from seawater or slightly salty water.

effluent A discharge of waste; pollutant.

eutrophication The overnourishment of aquatic ecosystems with plant nutrients (primarily phosphates and nitrogen), most often caused by agricultural **runoff**, urbanization, and sewage discharge, that results in a shortage of oxygen in the water, killing marine organisms.

groundwater Water beneath the earth's surface flowing slowly between saturated soil and rock that supplies wells and springs; held in underground reservoirs called **aquifers**.

nonpoint source The discharge of pollutants into water from a large, nondistinct area, such as runoff from farmlands and urban areas.

phytoplankton Microscopic floating aquatic plants found in the ocean.

point source A single identifiable source, such as a sewer pipe, that discharges pollution into water.

runoff   Water from precipitation and melting ice that flows over the earth's surface into streams, rivers, lakes, reservoirs, and wetlands, sometimes taking with it pollutants or dissolved substances, such as fertilizers or oil and salt from city streets.

salinization   The accumulation of salts in agricultural fields that eventually makes the soil unable to support crop growth.

subsidence   The sinking of the earth's crust as a result of removal of the groundwater supporting it.

surface water   Precipitation that does not seep into the earth or return to the atmosphere; found in lakes, ponds, and rivers.

transpiration   Process whereby water is transferred from a living plant to the atmosphere through water vapor emitted from exposed parts of the plant.

watershed   The land area draining into a river, river system, or body of water, bringing with it dissolved substances and sediment.

water table   The top surface of an underground aquifer.

INDEX

Acid rain, 16, 17, 20, 91
Africa, 17
Agricultural runoff, 45, 46, 47–51,
    52, 53–54, 57, 73, 75
Alaska, 15, 62
Amu Darya River, 20
Appalachia, 32
Aquifers, 18, 19, 20, 25, 27
Aral Sea, 20
Arcata, California, 78–79
Arizona, 53, 54, 86
Asia, 17
Aswân High Dam, 39
Audubon Society, 43, 79

Baikal, Lake, 17
Baikal nerpa, 17
Baltic Sea, 17, 75, 91
Basel, Switzerland, 60
Baton Rouge, Louisiana, 66
Beluga whales, 37
Benzene, 66
Bethlehem Steel, 76
Biomagnification, 49
Bioremediation, 63–65
Boston Harbor, 71
Boundary Waters Treaty, 86
Brine Shrimp, 41
Bureau of Reclamation, U.S., 51
Bullrushes, 78

Cadiz, 62
Cairo, Egypt, 39

California, 35, 41, 43, 49, 53, 54,
    55, 78, 79, 95
California, Gulf of, 54
California gulls, 41
Canada, 37, 47, 86
Carbon dioxide, 22
Carcinogens, 90
Center for the Protection and
    Restoration of Waters, 79
Central Arizona Project, 54
Chemical Corridor, 66
Chemical industry, 59, 60
Chesapeake Bay, 74, 75–76, 84
China, 31
Chlorine bleach, 21
Cholera, 18
Clarksdale, Mississippi, 66
Clean Water Act, 83
Cleo Wrap, 67
Cleveland, Ohio, 14
Coalition for Environmentally
    Responsible Economies
    (CERES), 67
Coastal wetlands, 27
Coast Guard, U.S., 63
Colorado River, 53, 54, 86
Comprehensive Environmental
    Response, Compensation, and
    Liability Act of 1980 (CERCLA),
    85, 86
Congress, U.S., 15, 84
Cubatao River, 93
Cuyahoga River, 14, 93

Cyanide, 90
Czechoslovakia, 17

Deforestation, 21, 36
Deicing compounds, 73
Desalinization plants, 94
Dioxin, 21
District of Columbia, 75, 84
Droughts, 33
Dysentery, 18

Earthscan, 16
Eastern Europe, 17, 90
Electroplating industry, 66–67
Environmental Protection Agency
    (EPA), 15, 46, 48, 66, 84, 91
Erie, Lake, 50, 88, 93
Estuaries, 27, 28
Euphrates River, 33
European Community, 91
Eutrophication, 36–37, 49–50
Exclusive Economic Zones (EEZs),
    89
Exxon, 15, 61, 62, 63, 67

Farm Act of 1985, 47
Federal Water Pollution Control
    Act, 83–84
Fertilizers, 15, 35, 40, 45, 46, 47,
    50, 56, 57, 73, 87
Florida, 49, 53
Florida Everglades, 46, 47
Food processing industry, 59
Fossil fuels, 20, 38

Genoa, Wisconsin, 65
Georges Bank, 88
Germany, 31, 91
Glaciers, 14, 94

Global warming, 21–22
Gorbachev, Mikhail, 17
Government, Canadian, 88
Government, U.S., 15, 54, 88
Gray water, 94
Great Britain, 17, 75
Great Lakes, 19, 86, 87, 88
Great Lakes Water Quality
    Agreement, 88
Groundwater aquifers, 19, 27,
    28–29, 35, 47, 48, 49, 50,
    52–53, 54
Gujarat, India, 38
Gulf of Maine Council on the
    Marine Environment, 88

Hazardous Waste Trust Fund. *See*
    Superfund
Heavy metals, 59, 73
Hepatitis, 18
Herbicides, 35, 47, 60, 93
Humboldt Bay, 78
Hydroelectric power, 13, 38–41
Hydrologic cycle, 25–26, 28

India, 51
Inland wetlands, 27, 28
Integrated pest management (IPM),
    57
International Convention for the
    Prevention of Pollution from
    Ships (MARPOL), 89
Irrigation, 29, 30, 35, 45, 50–52,
    54–55, 94
Israel, 33, 53

Japan, 31
Jordan, 53
Jordan River, 33

Kansas, 19
Kelco, 95
Kuwait, 33

La Crosse Boiling Water Reactor,
    66
Landfill, 20, 21
Latin America, 17, 32
Law of the Sea Convention, 89
Leachate, 20, 35, 59, 72, 93
Lead, 90
Leaking underground storage tanks
    (LUSTS), 15
Leaking Underground Storage
    Tank Trust Fund, 15
Lima, Peru, 80
Livestock, 49
London Dumping Convention, 89,
    90
Long Island, New York, 48
Los Angeles, California, 41, 43
Los Angeles Aqueduct, 42
Los Angeles Department of Water
    and Power, 41, 43
Louisiana delta, 65
Love Canal, 85
Low-flow toilets, 81
Low-level radioactive wastes, 90

Mackerel, 40
Maipo River basin, 60
Maine, Gulf of, 88
Malaria, 18
Maryland, 75, 84
Medical waste, 15
Mediterranean Action Plan, 90
Mediterranean Sea, 75
Metropolitan Water District of
    Southern California, 55
Mexico, 53, 54, 86

Mexico, Gulf of, 74
Microirrigation, 52
Middle East, 33
Mine runoff, 17
Mississippi River, 65, 66
Mono Lake, 41–43
Monongahela River, 14
Monsanto Corporation, 66
Municipal landfills, 72

Nasser, Lake, 40
National Coastal Zone
    Management Act, 84
National Environmental Policy Act
    (NEPA), 83
National Oceanic and
    Atmospheric Administration
    (NOAA), 73, 84
National Research Council, U.S.,
    87
New Orleans, Louisiana, 66
Newsweek, 65
New York City, 81
New York harbor, 74
Niagara Falls, New York, 85–86
Nile River, 39, 40
Nitrogen compounds, 49
Nonbiodegradable detergents, 14
North Sea, 37, 60, 61
Norway, 16, 17

Ocean Dumping Ban Act, 84
Ocean pollution, 60, 74
Oceans, 26–27, 28
Ogallala aquifer, 19, 52, 53
Oil dumping, 89
Oil spills, 15, 37, 61–63, 74
    cleanups, 62–65
Okeechobee, Lake, 47
Ontario, Lake, 88

Organic wastes, 47
Overfishing, 36

Paper recycling, 21
Paradichlorobenzene, 66
PCBs (polychlorinated biphenyls),
    60–61
Persian Gulf, 62
Pesticides, 15, 35, 45, 46, 47, 48,
    49, 56, 57, 60, 73, 77, 90, 93
Petrochemical industry, 59
Petroleum industry, 61
Phoenix, Arizona, 54
Phosphates, 49, 75, 77
Phytoplankton, 22
Pioneer Metal Finishing, 67
Pittsburgh, Pennsylvania, 14
Plastics, 36, 74
Poland, 17, 91
Polar ice caps, 14
Polychlorinated biphenyls. See
    PCBs
Potassium, 49
Power plants, 71, 72
Prince William Sound, 15, 62, 65
Providence, Rhode Island, 79
Puget Sound, 74
Pulp-and-paper industry, 17, 59, 60

Red Sea, 62
Regional Seas Programme, 90
Reservoirs, 18, 27, 28
Rhine River, 37, 60, 67
Rocky Mountains, 53, 54
Roemer, Buddy, 66
Royal Society of Canada, 87
Rubber industry, 60
Rush Creek, 42–43

Sacramento River, 54

St. Lawrence River, 37
Salinization, 35, 40, 50–51, 94
Salmon, 38
Salt, 73
San Diego, California, 95
Sandoz Corporation, 60
San Francisco Bay, 54, 74
San Joaquin Valley, 35
San Juan de Miraflores, Peru, 80
Santa Monica Bay, 74
Sardar Sarovar Dam, 38
Sardines, 40
Saudi Arabia, 94
Sauget, Illinois, 66
Scandinavia, 17, 91
Schistosomiasis, 18, 40
Seals, 17, 37
Sewage, 37, 40, 46, 74, 76, 90
Sewage treatment plants, 15, 18,
    71, 73, 75, 76, 84
    alternative designs, 78–80
    stages of treatment, 76–78
    and storm runoff, 71
Shrimp, 40
Siberia, 17
Sinkholes, 53
Solar aquatics, 79
Soviet Union, 17, 20, 31, 59
Storm drains, 71, 84
Streams, 27, 28
Subsidence, 35, 53
Suburban runoff, 73, 75
Superfund, 85, 86
Sweden, 16, 17, 67, 91
Syr Darya River, 20

Temik, 48–49
Texas, 19, 53
Textile industry, 59
Thermal pollution, 72

Third United Nations Conference
on the Law of the Sea, 89
3M Corporation, 67
Tigris River, 33
Toxic chemicals, 37
Toxic waste dumps, 59
Tucson, Arizona, 54
Typhoid fever, 18

Underground reservoirs. *See*
Groundwater aquifers
United Nations, 16
United Nations Economic
Commission for Latin America
and the Caribbean, 32
United Nations Environmental
Programme (UNEP), 90
United States, 15, 19, 30, 31, 45,
47, 49, 54, 86
Urban runoff, 73, 75

Valdez, 15, 61, 62, 63, 65, 67. *See
also* Oil spills
Valdez Principles, 67–69

Washington, 84
Waste reduction, 36
Wastewater, 66, 76
pretreatment, 66–67
Water conservation legislation, 83,
84
Water hyacinths, 78
Waterlogging, 50–51, 94
Water pollution

agricultural, 36, 45, 46, 47–52
and disease, 17–18
industrial, 36, 37, 59–62,
66–67, 73, 74
legislation regulating, 15, 16,
60, 62–63, 83–84, 85–86,
88, 89–90
municipal, 35, 71–73, 74, 75
transnational, 15
Water recycling, 36
Water Resources Council, 45
Water rights, 55
conflicts over, 33
international law governing,
86
trading of, 55
Water use
agricultural, 30, 31, 35, 45–57
conservation techniques, 93,
95, 96–97
domestic, 30
industrial, 30, 31, 59
management of resources, 20
measurement, 29
municipal, 30, 31, 71
Western Europe, 17
Wetlands, 46
World Health Organization
(WHO), 32
World Resources Institute, 31

Yellow fever, 18
Yuma Desalting Plant, 54, 86

(From U.S./English system units to metric system units)

Length

1 inch = 2.54 centimeters
1 foot = 0.305 meters
1 yard = 0.91 meters
1 statute mile = 1.6 kilometers (km.)

Area

1 square yard = 0.84 square meters
1 acre = 0.405 hectares
1 square mile = 2.59 square km.

Liquid Measure

1 fluid ounce = 0.03 liters
1 pint (U.S.) = 0.47 liters
1 quart (U.S.) = 0.95 liters
1 gallon (U.S.) = 3.78 liters

Weight and Mass

1 ounce = 28.35 grams
1 pound = 0.45 kilograms
1 ton = 0.91 metric tons

Temperature

1 degree Fahrenheit = 0.56 degrees Celsius or centigrade, but to convert from actual Fahrenheit scale measurements to Celsius, subtract 32 from the Fahrenheit reading, multiply the result by 5, and then divide by 9. For example, to convert 212° F to Celsius:

$$212 - 32 = 180 \times 5 = 900 \div 9 = 100° C$$

PICTURE CREDITS

Agricultural Research Service, USDA: pp. 12, 44, 48, 51, 56, 82, 85, 87, 90, 92; AP/Wide World Photos: pp. 19, 24, 34, 39, 42, 64, 80, 98; The Bettmann Archive: pp. 58, 61; © Greenpeace/Morgan: p. 70; Reuters/Bettmann Archive: p. 16; UPI/Bettmann Archive: pp. 68, 74, 95; USDA: p. 29

ABOUT THE AUTHOR

KAREN BARSS is a writer and editor from Boston, Massachusetts. She is currently coordinator of print projects for public television station WGBH, where she develops educational materials for teachers and students to accompany national broadcasts. She was project director for the high school environmental activity guide that accompanied the 10-part series *Race to Save the Planet* and is now developing print materials for an interactive videodisc version of the series for middle schools. Ms. Barss received her B.A. in European history from Middlebury College.

ABOUT THE EDITOR

RUSSELL E. TRAIN, currently chairman of the board of directors of the World Wildlife Fund and The Conservation Foundation, has had a long and distinguished career of government service under three presidents. In 1957 President Eisenhower appointed him a judge of the United States Tax Court. He served Lyndon Johnson on the National Water Commission. Under Richard Nixon he became under secretary of the Interior and, in 1970, first chairman of the Council on Environmental Quality. From 1973 to 1977 he served as administrator of the Environmental Protection Agency. Train is also a trustee or director of the African Wildlife Foundation; the Alliance to Save Energy; the American Conservation Association; Citizens for Ocean Law; Clean Sites, Inc.; the Elizabeth Haub Foundation; the King Mahendra Trust for Nature Conservation (Nepal); Resources for the Future; the Rockefeller Brothers Fund; the Scientists' Institute for Public Information; the World Resources Institute; and Union Carbide and Applied Energy Services, Inc. Train is a graduate of Princeton and Columbia Universities, a veteran of World War II, and currently resides in the District of Columbia.